THINK SMART

Also by Richard Restak, M.D.

The Naked Brain

Poe's Heart and the Mountain Climber

Mysteries of the Mind

Mozart's Brain and the Fighter Pilot

The New Brain

The Secret Life of the Brain

The Longevity Strategy (with David Mahoney)

Older and Wiser

Brainscapes

The Modular Brain

Receptors

The Brain Has a Mind of Its Own

The Mind

The Infant Mind

The Brain

The Self Seekers

The Brain: The Last Frontier

Premeditated Man

THINK SMART

A Neuroscientist's Prescription
for Improving Your Brain's Performance

―――――――――

RICHARD RESTAK, M.D.

Riverhead Books
a member of
Penguin Group (USA) Inc.
New York
2009

RIVERHEAD BOOKS
Published by the Penguin Group
Penguin Group (USA) Inc., 375 Hudson Street, New York, New York 10014, USA • Penguin Group
(Canada), 90 Eglinton Avenue East, Suite 700, Toronto, Ontario M4P 2Y3, Canada (a division of
Pearson Canada Inc.) • Penguin Books Ltd, 80 Strand, London WC2R 0RL, England • Penguin Ireland,
25 St Stephen's Green, Dublin 2, Ireland (a division of Penguin Books Ltd) • Penguin Group (Australia),
250 Camberwell Road, Camberwell, Victoria 3124, Australia • (a division of Pearson Australia Group Pty
Ltd) • Penguin Books India Pvt Ltd, 11 Community Centre, Panchsheel Park, New Delhi—110 017,
India • Penguin Group (NZ), 67 Apollo Drive, Rosedale, North Shore 0632, New Zealand
(a division of Pearson New Zealand Ltd) • Penguin Books (South Africa) (Pty) Ltd,
24 Sturdee Avenue, Rosebank, Johannesburg 2196, South Africa

Penguin Books Ltd, Registered Offices: 80 Strand, London WC2R 0RL, England

The drawing on page 185 is by Ashley Tucker.
The chart on page 211 is reproduced with permission of the
Society for Human Resource Management (SHRM).

Library of Congress Cataloging-in-Publication Data

Restak, Richard M., date.
Think smart: a neuroscientist's prescription for improving your brain's performance /
Richard Restak.
p.   cm.
Includes bibliographical references and index.
ISBN 978-1-59448-873-3
1. Brain—Aging—Prevention.   2. Mental work.   3. Nutrition.   4. Mind and body.
5. Self-care, Health.   I. Title.
QP398.R47   2009   2009002517
612.8'2—dc22

Printed in the United States of America
1   3   5   7   9   10   8   6   4   2

Book design by Michelle McMillian

To my brother, Christopher

CONTENTS

To remain mentally sharp, you have to deal with familiar things in novel ways. But most important of all, you have to have a sense of curiosity. If interest and curiosity stop coming automatically to you, then you're in trouble, no matter how young or old you are.

—*Art Buchwald*

INTRODUCTION

"What should I do to keep my brain working at its best?"

I'm frequently asked that question. It makes sense as I've written eighteen books about the human brain. But I recently decided to undertake a personal odyssey aimed at discovering what I can do to improve my brain. This seemed especially important, since I will soon be at an age when brain function typically declines unless deliberate steps are taken to maintain it.

I come to this task from a unique vantage point, based on my experience as both a neurologist and an author. Over the years, I've become acquainted with many of the world's foremost neuroscientists (brain researchers). I've talked with them during neuroscience meetings, observed them in their laboratories, and read their published writings in which they explain their discoveries. What could be more natural, I wondered, than to ask these brain scientists, the best in their field, "What are *you* doing to keep your brain functioning at its best?"

Their answers often surprised me. And I suspect they will surprise you.

Using their answers, coupled with my own work in cutting-edge brain research, I've set out in these pages my personal program to improve your brain's functioning. It can be used by anyone interested in developing and maintaining an optimally functioning brain. Whether you are young or old, rich or poor, male or female, these insights will help your brain to be more efficient, more effective, and more engaged.

PART ONE

Discovering the Brain

—————

When I was a medical student, neither teachers nor students placed much emphasis on the brain. The curriculum included only a first-year course in neuroanatomy, followed two years later by a one-month rotation spent working with patients on the neurology wards. After graduation, most medical students tended to avoid specialty training in careers devoted to treating the brain (neurology and neurosurgery), based on the general perception that not much could be done to heal or even improve the lives of many of the patients afflicted by brain diseases. I remember vividly my father's disappointment when I told him I was interested in neurology and psychiatry rather than obstetrics (his specialty). "You can't do anything for most of the patients you'll encounter in either of those specialties, and it's awfully depressing to just diagnose and not treat," he told me.

A lot has changed since then. We have learned more about the brain in the past decade than we did in the previous two hundred years. If he were still alive, my father would be amazed at the effective treatments now available for many brain diseases such as multiple sclerosis, migraine, and epilepsy, to mention just the most common. Neuroscience (brain science) is currently one of the most popular career choices among students attracted to science. Psychiatry and neurology are in the process of merging into the hybrid discipline of neuropsychiatry. But these advances didn't happen spontaneously. The advance from nihilism and pessimism toward curiosity and hope was stimulated principally by new ways of imaging the brain.

Until the middle part of the twentieth century, what little was known about the brain consisted of a mélange of speculation and dogmatism based on hosts of hoary old men staring through microscopes at brightly colored dye-stained neurons. Thanks to advances in brain-imaging techniques over the last thirty-plus years, it's currently possible for neuroscientists (many of whom are now women) to observe the development of the brain in real time and without any need for either speculation or dogmatism. The principle behind these illuminating (in both senses of the term) imaging techniques is straightforward: blood flow to the brain varies with activity. The greater the activity, the greater the flow of blood needed to replenish the oxygen and glucose used by the active neurons. This isn't any different from what happens elsewhere in the body.

When you lift a hundred-pound barbell at the health club in an effort to attract the attention of someone nearby in whom you're romantically interested, blood flow increases in the muscles of your arms and chest based on the increased need by those muscles for oxygen and glucose. Similarly, when you use a specific circuit in the brain, the components of that circuit will become more active and call on the circulatory system to provide additional glucose and oxygen. Positron-emission tomography (PET) and functional magnetic resonance imaging (fMRI) detect the changes in blood flow within active parts of the brain and record them while the subject lies within a special scanner.

Thanks to fMRI imaging and other techniques, we know that the brain never wears out; it gets better the more we use it; it changes in structure and function throughout our lives. As a consequence of this *plasticity* we sculpt our brains according to our life experiences. As a result, no two brains are exactly alike, not even the brains of identical twins who, while they share the same genetic makeup, don't share identical experiences. Due to this diversity in the brain's organization and structure from one person to another, it's often possible to reach valid conclusions about a person on the basis of his brain's organization.

For instance, while looking at an fMRI, a trained observer can distinguish the brain of a skilled pianist from that of a nonpianist. That trained observer will note that the pianist's brain shows increased activation in the finger areas of the motor cortex while she listens to a piano concerto. The same thing happens if she just watches someone playing any musical composition on the piano. But it won't happen if she observes that person merely making random finger movements on the keyboard: under these circumstances, the pianist's brain responses won't differ from those of a person with no special musical expertise or interest. A similar specialization occurs in dancers. A ballet dancer will show greater brain activation while he watches other ballet dancers perform. This won't happen if he watches ballroom dancers.

Nor is brain specialization limited to the arts. If that pianist at the conclusion of her performance takes a cab from the concert hall to her apartment, her cabdriver's brain is likely to have an enlarged hippocampus—a brain area heavily involved in spatial visualization and navigation. The same is true for any specialized occupation: A surgeon will show greater activation in the hand area of the motor and sensory cortex than will a doctor who doesn't perform surgery.

The pianist, the ballet dancer, the taxi driver, and the surgeon have shaped their brains by virtue of their experiences. The same thing is true for all of us. We create new patterns of neuronal organization according to what we see, what we do, what we imagine, and most of all, what we learn. Learning something new involves establishing a pathway within the brain made up of millions of brain cells. As we learn more, these pathways increase in complexity—a process similar to the branching of a tree as it grows.

Thanks to its plasticity, the brain can be thought of as a tree of knowledge. When in full bloom, a tree blossoms: roots give off branches, twigs, and leaves. Similarly, learning increases interaction within the brain with more and more other neurons establishing fuller and richer circuits. But if learning stops, the brain, like a tree losing the luxuriant structure seen at full bloom, reverts to a state corresponding to that of a tree in winter.

You can picture human brain development as a continuum ranging from infant-child to adolescent to adult and, finally, to the mature older brain. Each stage of development along this continuum calls for specific approaches to brain enhancement. Equally important, lessons learned at one stage of brain development can be usefully applied at every other level, starting from its earliest inception until its eventual dissolution and demise in old age. Thus knowing principles drawn from the study of the infant brain will help you enhance your adult brain.

Indeed, infant and adult brains share many of the same challenges: stimulation but not overstimulation, maximizing plasticity, establishing and maintaining nerve cell (neuronal) circuits in the face of a steadily decreasing loss of neurons, among others. That last point (fewer cells but greater connections) may strike you as strange, even paradoxical, as it did me when I first learned about it during my neuroscience training. Indeed, this "improved function with fewer components" principle is one of the great paradoxes of the human brain.

At birth our brain possesses almost all of the neurons it will ever have. Maximum brain cell number is achieved during an explosive growth period that takes place between the third and sixth months of life. During the next three months, before birth and extending into the first two years of life outside the womb, the total number of neurons decreases, while the functional connections among the surviving neurons, the synapses, increase. Thus the newborn infant is equipped with considerably fewer neurons than it possessed in the womb but a far greater number than it will have as an adult.

Now here's the paradox: As we progress from infancy to childhood to adolescence to adulthood, the brain's performance improves and yet does so with fewer neurons. No mechanical device operates with greater efficiency as its components are gradually taken away. Imagine removing parts of your car's engine every year and thereby improving its performance. A similar situation exists in regard to every part of the human body except the brain: remove healthy heart, lung, liver tissue, and you wind up with a compromised organ.

Enriching the Brain

Only recently have neuroscientists been able to account for this odd state of affairs whereby we have more brain cells during the period when we're learning to say "Mama" and "Dada" than when we're learning geometry or later heading up a Fortune 500 company.

In order for the brain to develop normally, large numbers of brain cells must first be generated (a period referred to as proliferation) and many of them later eliminated (a process neuroscientists refer to as pruning). Pruning results in fewer but faster and more effective brain cell connections (synapses). An estimated 40 percent of synapses generated during infancy are eliminated by adulthood. "Use it or lose it" is the operative term that describes this process—and it applies across the entire life span of the brain.

Whether we're in the bassinet or in the boardroom, those brain cells that establish connections with other cells will be maintained; those that fail to link with others will die off. A similar principle—dubbed "neural Darwinism" by the Nobel Prize–winning neuroscientist Gerald Edelman—applies at the level of brain cell circuits (networks). Those brain circuits that are actively maintained and challenged will endure and grow stronger; those that are used infrequently, if at all, will gradually disappear—sort of like friendships.

As part of this process of challenge and maintenance, novelty and enriched experiences work like fertilizer on brain growth and development. We know this from a series of now famous experiments comparing the brains of two groups of caged rats, which were carried out in the 1970s by neuroscientist Bill Greenough, then at the University of Illinois at Champaign–Urbana.

One group of rats lived alone in the equivalent of "lockdown" (no companions, nothing to do, etc.). The second group was treated more like white-collar criminals given the opportunity to spend their time in comparatively plush surroundings (at least by lab rat standards). Their cages were fancier and furnished with wheels to spin, ladders to scale, and other rats to play with—"the rat equivalent of Disneyland," as Greenough characterized it. These more favorably endowed rats became more physically and socially active—networking and coexisting with other rats in the kinds of competitive though generally peaceful projects possible for a rat spending its days and nights in an animal lab in Illinois.

When studied under the microscope, the brains of the rats raised in the "enriched" environments contained 25 percent more synapses per neuron than those of the isolated rats. This increase in synapses translated into cleverer rats that were quicker to wend their way through mazes and learn landmarks faster. The message from the Greenough experiments seemed fairly straightforward: If you want a rat to grow up smart instead of stupid, make its life more challenging; increase the rat's opportunities for sensory stimulation, physical exercise, and socialization. Each of these factors increases blood supply to the brain, enhances brain development, and leads to the creation of smarter rats.

When I first learned about Greenough's research, I was impressed but skeptical. Shouldn't his findings be placed within the context of the normal life of a rat? Even the rats in the enriched-environment cages lived incredibly impoverished lives compared with their wild cousins that live in sewers and back alleys richly supplied with complicated mazes, tunnels, and debris—to say nothing about the huge numbers of other rats that must be dealt with. From that vantage point, it's fair to say that *all* of the rats in Greenough's experiments lived environmentally impoverished lives. Here's what I think is a reasonable summary of his enrichment research: The rat living in an enriched laboratory environment winds up with a greater number of synapses and more enhanced brain development than a rat living in anything other than its natural environment outside a laboratory.

Having established the value of novelty and an enriched environment as stimulants for brain development in rats, Greenough's research prompted a tantalizing question: Would environmental enrichment lead to enhancement in the human brain? Obviously, a similar experiment could not be carried out in humans. But even though the supporting evidence is less direct, it's nonetheless persuasive. Not only do infants raised in institutions show stunted intellectual and social development compared with other infants transferred from the institution to an adoptive family, but their brains also have fewer connections linking different parts of the cortex. And children placed in "high quality" day care (lots of toys, interaction with other kids, and dedicated resourceful teachers) go on to perform better in elementary school than children from centers where the emphasis is on supervision and control. Nor does the influence of social and cultural enrichment on brain performance end in childhood; it continues throughout the life span. Education, both formal and informal, and practical experience are the greatest environmental enrichment agents. Thanks to new imaging techniques, it's possible to see the brain changes induced by learning and experience. For example, among London cabdrivers, those with the most experience in successfully navigating that city's labyrinth of streets show the most significant enlargement of the hippocampus—a sea horse–shaped structure known to be important in spatial learning and memory. Other imaging studies demonstrate that, in general, life experiences leading to the development of special abilities (musical, athletic, artistic) also induce structural changes in the brain areas that mediate these abilities.

Maturing the Brain

When does the brain reach maturity? That depends on your definition of *maturity*. From the behavioral point of view we can all bring to mind people who never seem to mature. Throughout their lives they continue to grapple with issues involving authority, identity, and self-assertion (among others)—issues that the majority of people resolve before casting their first vote. But if we talk about maturity from the point of brain structure and function, the story is quite different.

Before the 1970s it was widely believed that all of the different regions of the brain developed at same time. But in the 1980s this belief was found not to be true. Brain regions that control primary functions such as movement, seeing, and hearing develop first, followed by areas concerned with language and thinking. The last brain regions to mature are the prefrontal and temporal areas, which integrate attention, language, and decision making. This sequence of development is mirrored behaviorally: burps precede elocution: the infant sees and hears prior to speaking or learning words and concepts.

This insight into the sequential development of the various parts of the brain resulted from revolutionary imaging devices such as magnetic resonance imaging (MRI), which provides a window on brain structure—the geography of the brain—coupled with functional MRI, or fMRI, which shows color-coded pictures of ongoing brain activity. Both imaging devices reveal striking differences between the child and adolescent brain and its adult counterpart.

Total brain volume reaches a peak at about eleven years in girls and fifteen years in boys, and is followed by a slow decline over the adult years. The most striking developmental change occurs in exponential growth within the frontal lobes.

Located farthest to the front of the brain, the frontal lobes are responsible for our most evolved feelings and behaviors such as ethics, altruism, and compassion. The frontal lobes are also important in foresight, planning, and follow-through. Foreseeing the likely consequences of one's actions requires normally functioning frontal lobes. Some adults seem to be frontally challenged when it comes to these frontal-lobe functions.

Since the frontal lobes develop at a much slower pace than other brain areas, the humanizing qualities mediated by the frontal lobes are in scant supply early in life. Spend a few minutes in a playground and you can observe that toddlers and very young children need to be reminded to share, to avoid hurting other children's feelings, and to settle disputes without recourse to verbal or physical attacks. A similar need for externally imposed structure in the absence of internal controls occurs among some adults. Although I've never committed a crime, my work as a forensic neuropsychiatrist has taken me behind the walls of more prisons than I care to count. I've found that many prisoners, especially those serving time for violent crimes, suffer from deficiencies in frontal lobe function. They can't plan their lives or control either their emotions or their behavior. In some cases, I can demonstrate these frontal lobe deficiencies through testing and imaging. This can prove helpful by providing a partial explanation for the actions that led to the prisoner's incarceration. In other instances, the studies are normal by the criteria of currently available technology.

For those of us fortunate enough to possess brains with frontal lobes that underwent normal maturation, the process began in adolescence. As the frontal lobes begin to mature, each neuron becomes a component in any number of vast interconnected networks. Just as a person can simultaneously participate in many networks (work, church, local community, clubs), so too the individual neuron may participate in multiple circuits and networks within the brain. This social analogy is a good one for understanding the brain throughout its life cycle. Just as the totally isolated human being operates at great disadvantage, a neuron that fails to establish connections with other neurons in the brain fails to thrive and eventually dies off. The richness and complexity of an individual brain depends on the networking made possible by millions of interconnected neurons linked together to form untold numbers of circuits. This holds true whatever the age of the brain.

The Adolescent Brain

As the neurons and their connections become increasingly networked during adolescence, more evolved feelings and behaviors first begin to express themselves, sometimes to excess—the uncompromising, sometimes exasperating idealism of the adolescent, for example. In time, frontal lobe function becomes more balanced, with youthful idealism coexisting alongside the recognition of practical realities. But this balance isn't fully achieved until well into adulthood.

Much of the "immature" behavior that is so characteristic of the teen years (trouble foreseeing consequences, bad judgment, impulsiveness, and difficulty controlling impulses—especially violent and sexual ones) results from immaturity of the adolescent prefrontal cortex. The good news is that if you wait long enough and are patient enough, judgment, self-control, and other frontal lobe functions will improve.

In short, if you interact with adolescents as parent or teacher, you will observe the behavioral correlates of frontal lobe immaturity followed by a slowly, painfully emerging maturity: the behavioral patterns mature as the adolescent brain matures.

One more point about the adolescent brain: It doesn't manage stress very well. Typically stress in an adult causes a rise in cortisol levels (a measure of stress) followed by a gradual decrease over an hour or two. In adolescents, that burst of cortisol hangs around a lot longer, resulting in sustained exposure of the brain to harmful effects, such as shrinkage of cells in the hippocampus (resulting in memory loss and depression) and the amygdala (resulting in anxiety and other overwhelming emotions). This has important consequences because the hippocampus, the amygdala, and the prefrontal cortex are the three brain areas that undergo major changes during adolescence. If these brain areas are damaged by stress hormones, the effect can extend into adulthood—the basis for observations among mental health specialists linking adolescent stress to adult behavioral and emotional problems. And while the stress-induced changes in the hippocampus often improve when the stress is reduced, the changes in the amygdala—emotional changes—don't always change back.

So don't be too hard on adolescents. Their brains must successfully respond to dual challenges. Because of the internal reorganization that is taking place between the ages of ten and fifteen, the adolescent brain is singularly adept at learning. Yet learning is made more difficult thanks to the immaturity of the frontal lobes. As a parent or teacher it's helpful to keep in mind that the adolescent's failures in concentration, focus, motivation, and consistent effort result not from willfulness or laziness or, God forbid, "stupidity," but from poor integration of the frontal lobes.

Fortunately, parents, teachers, and others can counteract these tendencies by helping teenagers take an active role in determining the structure and functioning of their brain. If the adolescent is encouraged to concentrate on music, math, or sports, for instance, the brain will incorporate these activities in the form of neuronal circuits. If the teenager, in contrast, spends the day "hanging out" or mindlessly gossiping on a cell phone, the brain will fashion circuits for these activities as well. In essence, adolescents choose the brain cells and circuits that will survive on the basis of the activities they engage in.

The Adult Brain

With the completion of adolescence the brain is fully functional and, barring disease or injury, will remain so until late in life. The transition from adolescence to adulthood involves:

The dominance of the frontal lobes over the limbic system. As a result, the adult, in contrast to the adolescent, is no longer as dominated by impulses, moods, or emotions in general. This comes about because of several frontal-lobe–mediated changes that increase ability to put things in context; to plan and organize one's life; to guard against distractions caused by thoughts, impulses, and fantasies; and to focus on the present while simultaneously envisioning how one can change current circumstances by judicious effort.

Specialization within the brain, thanks to the creation of millions of new circuits, which vary from one person to another based on life experience. The most dramatic specialization takes place within the two hemispheres. The brain of a lawyer differs from that of an architect, for instance, thanks to enhancement of different circuits within the two hemispheres. The lawyer's facility in the use of written and spoken language relies heavily on the circuitry of the left hemisphere, where language is primarily processed; the architect's ability to envision structures in three-dimensional space results from the buildup of circuits within the right hemisphere. But such arrangements aren't mutually exclusive. Thanks to the brain's plasticity, circuitry within both sides of the brain can develop equally. (My brother is both an architect and a lawyer.)

Continuation of the brain's plasticity. Think of the brain as almost infinitely plastic throughout our lives. I've qualified my statement with *almost* because of the influence of genes on the brain's structure and function. And although this brain plasticity diminishes with age, it never completely disappears. Until the day we die our brain remains capable of change, according to the challenges that we set for it.

The brain reaches its maximum size (measured by weight) in early adult life and decreases by about 10 percent over an average life span. And what a marvelous organ it is.

The adult human brain weighs about three pounds and contains about a hundred billion brain cells (neurons) with a million billion connections (synapses) linking those neurons to each other.

While it's hard to grasp the scope of a number like one million billion, it's easier if you approach it in stages. One million is easy to conceptualize: 1,000 x 1,000, about the population of a moderate-size city. A billion is one thousand times one million—a thousand small cities or a hundred very large cities. To put one million billion into perspective, there are about six billion people currently living on our planet—a small number compared with those hundred billion nerve cells and a truly paltry number compared with those million billion connections.

Although the number of synapses generally decreases as a person ages, most neuroscientists no longer believe that brain cell loss occurs at the frequently quoted figure of fifty thousand cells a day. Nor is the loss of neurons equally distributed throughout the brain. Rather, recent evidence points to a more selective loss of neurons with little or no significant neuron loss in many cortical regions used in normal cognition. Most striking is the variability in brain aging between one older person and another. This is undoubtedly secondary to genetic factors. As one neuroscientist humorously put it, "The best way to guarantee a normal brain in old age is to pick your parents carefully."

Even more important than loss of neurons and the thinning of synaptic connections, especially in the frontal lobes, that occurs as we age, is the loss of cells from clusters of cells (nuclei) about the size of a pinhead located in the brain stem, which itself is only about the length of an adult forefinger. Dubbed "juice machines" by neuroscientist Paul Coleman, these clusters of cells send ascending fanlike projections to many parts of the cortex. The brain's chemical messengers (neurotransmitters) travel along these projections. Reduction in the levels of neurotransmitters leads to many of the infirmities that increasingly afflict us as we get older: memory loss, depression, decrease in overall mental sharpness, inefficient information processing. Fortunately, these infirmities can be improved by drugs that rev up or supplement deficient neurotransmitters.

Finally, as we age, our brain accumulates chemical break-down products resulting from the sum total of all of the metabolic processes that take place within our brain cells over our life span. Every older person's brain contains a certain quantity of these products (but nowhere near the amount seen in people suffering from Alzheimer's disease). Further, while some loss of neurons occurs normally with aging, the loss can be compensated by increases in the networking capacity of the remaining neurons. While the number of neurons decreases from birth onward, fewer but stronger and more enduring connections form among the remaining neurons. This capacity to compensate for the loss of its components makes the brain the only known structure in the universe that works more efficiently despite a loss of its components. To this extent the brain is unique among both biological and mechanical structures: over the years it doesn't "wear out."

Given all this change within the brain as we progress from infant to child to adolescent to young and, finally, aged adult, here is the key question: What can be done to preserve and enhance the brain's powers? The next 230 pages suggest a practical program based on my own experience as doctor and author, and the suggestions of many of the world's most prestigious neuroscientists. Let's start with the most basic need of all: What does the brain like to eat?

PART TWO

Care and Feeding
of the Brain:
The Basics

In Search of the "Brain Diet"

It takes only a glance at the cover of self-improvement magazines to confirm our nation's enduring interest in discovering the perfect diet. Curious readers can also choose from a steady stream of books making various dietary claims. Some suggest that one should eat, for instance, more carbohydrates, or fewer carbohydrates, or perhaps no carbohydrates at all. But despite their differences, all of the books conform to a common theme: If you follow *this diet,* you will be thinner, healthier, and sexier than you've ever been before. But whether the diets originate in South Beach or Beverly Hills— and however successful they may be for inducing general weight loss in individual dieters—they offer scant advice for people more interested in preventing mental rather than physical flab: tuning up their frontal lobes rather than their abs.

One reason for this absence of a "brain diet," I've discovered from consulting experts on the brain and nutrition, is that the specific class of foods we eat may be less important to the health of our brain than the number of calories we consume. At least that is the message conveyed by researchers on brain health in animals. Sixty-five years of animal research confirms that in every animal in which it has been tried so far, caloric restriction slows the onset of degenerative diseases such as dementia, cancer, diabetes, and other illnesses associated with loss of brain function. Caloric restriction (defined as a balanced reduction of the protein, carbohydrate, and fat content of food without reduction of the nutrient content) also increases the life span. As a rough rule of thumb, an animal of whatever species that eats 35 percent fewer calories will live 35 percent longer.

If we consider animal research, should we adopt caloric restriction diets? We should if we want to lessen our chances of getting Alzheimer's disease, according to research on mice fed a diet containing 30 percent fewer calories than mice allowed to eat as much food as they chose. Examination of the brains of the mice on the calorie-restricted diet showed a reduction of amyloid, the brain-clogging protein implicated in Alzheimer's disease. Moreover, these mice outperformed their counterparts on an unrestricted caloric diet.

According to researchers at Oregon Health and Science University and the National Institute on Aging (NIA), caloric restriction, combined with periodic fasting, reduced the impairment of "Alzheimer's" mice. Specifically, these mice, with a progressive illness similar to many aspects of Alzheimer's disease, performed better on learning and memory tests and explored their surroundings with renewed vigor.

While I'm impressed with the amyloid-reducing properties of caloric restriction in mice, and even more impressed by the mice's enhanced mental performance, neither of these findings is likely to exert much of an influence on my planning for this evening's meal. Sure, I'll readily concede that I'm eating a tad more than I need; nonetheless, since I'm not a mouse, I'd like to base my decision about caloric restriction on research involving animals higher on the food chain. But such research isn't likely to be available anytime soon, because of life-span differences among species. Two or three generations in the mouse world doesn't come anywhere close, time-wise, to the more than two hundred years corresponding to three human generations.

But this extended time line hasn't stopped some people from intentionally going hungry in order to improve brain function and live longer. In one study sponsored by the National Institute on Aging, forty-eight men and women were randomly assigned to one of two diets. The first was sufficient to maintain current weight, while the second diet reduced caloric intake by 25 percent. After six months, those on the restricted diet had lower body temperatures and levels of insulin whenever they ate—two key characteristics of long-lived people and animals. In essence, the body's energy requirements decrease to meet the reduced number of calories in the diet—the explanation why people on caloric-restricted diets don't starve to death. The body simply readjusts to less food. All of which leaves unsettled whether a severely calorie-restricted diet makes sense.

The important concept—and this is drawn from both human and animal research—is that in order to tune up your brain and reduce your likelihood of Alzheimer's disease, you may not have to cut back drastically on your caloric intake but simply keep your caloric intake low enough to prevent obesity.

Control Your Weight

Nearly two-thirds of adults in the United States are overweight, with 30.5 percent considered obese. While everyone is aware of the general medical consequences of obesity (heart disease, hypertension, and diabetes), the effect of obesity on brain function has only recently been uncovered.

Mice placed on diets high in saturated fat and "empty calories" significantly underperform in tests of memory when compared with mice on a normal mouse diet. Accompanying this performance failure are all of the bad things associated with such a diet: higher levels of plasma triglycerides, total cholesterol, and both high- and low-density lipoprotein cholesterol. Tests involving reinforcing rewards are also affected, i.e., the obese mice require more time to learn to press a lever in order to obtain a squirt of milk or fruit juice.

"Our findings show that fast-food diets impair memory in mice and make their brains more vulnerable to toxins," says Veerendra K. Madala Halaggappa, of the National Institute on Aging. "If such diets have similar effects in humans, then reducing the amount of fat and empty calories may improve one's memory and increase resistance to age- and stress-related cognitive impairment."

Dr. Halaggappa's comment suggests that reducing the dietary fat and empty calories in our diets will improve our memory and brain function in general. Think about the potential benefits of that insight for the mental functioning of a nation with a penchant for fast foods. In addition, research over the past year points to elevated cholesterol and obesity as risk factors for the development of Alzheimer's disease.

In short, measures to control obesity are worthwhile, whatever effort may be required, since, as tests of cognitive performance in humans indicate, obesity is more often associated with cognitive impairment than with age, gender, education, or IQ. Particularly affected are the functions carried out primarily by those all-important frontal lobes. As mentioned earlier, these most developed brain areas are known to be associated with setting and keeping to goals, controlling impulses, and monitoring one's own behavior—three special problem areas for the chronically obese.

After deciding to reduce caloric intake, what other dietary measures are most likely to protect and enhance our mental functioning?

To begin with, follow the oldest dictum in public health: *Primum non nocere* (First do no harm). Start by identifying and eliminating from your diet those foods that can be proven to cause harm. The chief villains here are processed fats. Especially harmful are the so-called trans fats, formed when liquid oils are transformed into solid fats by adding hydrogen to vegetable oil.

About twenty years ago food manufacturers started adding trans fat to their processed foods to prolong shelf life and stabilize flavor. Thanks to trans fat, crackers, cookies, and snack foods will stay fresh for years. Once ingested, however, these hydrogenated fats clog arteries in the brain and heart, leading to cognitive decline (memory is especially affected) and heart attacks. The stiffer and harder the fat—which varies directly with the degree of hydrogenation—the greater the artery-clogging effect. Trans fats exert the same occluding effect on arteries as bacon grease exerts on the drain in the kitchen sink. In order to avoid all this, follow the following guidelines:

Avoid foods labeled "hydrogenated or partially hydro-

genated." Be especially wary of fast foods (fried chicken, bis-
cuits, fried fish sandwiches, french fries, and fried foods
in general), most store-bought cookies and crackers, potato
chips, margarine, doughnuts, and muffins. In general, foods
that come from nature (unprocessed) don't contain trans fats.
That's one of the reasons nutritionists encourage us to eat fruits
of all types, vegetables, chicken, whole-grain breads, and some
cereals. Fortunately, label reading has become a lot easier
since 2006, when the U.S. government started requiring that
trans fat be listed on food labels along with saturated fat and
cholesterol. Maintain a zero tolerance for any trans fats in
your diet.

What's Good for the Brain Is Good for the Heart

After eliminating the harmful foods, how do you decide
which foods will help brain function? One way of identifying
brain-enhancing foods is to select foods known to slow the
rate of cognitive decline in older people. Admittedly, the reasoning here isn't perfect ("Anything that slows the brain
changes associated with aging should be helpful in maintaining brain function earlier in the life span"), but that's about
as good as we can do until it is possible to directly observe the
effects of different foods on brain function.

Foods thought to slow the rate of cognitive decline
include vegetables, especially green leafy ones. Not only do
green leafy vegetables contain more vitamin E than other
vegetables, but they're also frequently consumed with added

fats (e.g., salad dressing, mayonnaise) that increase the absorption of vitamin E and other fat-soluble antioxidant nutrients.

Antioxidants are chemicals that interrupt oxidation, the process that causes things to decay. A good example of oxidation in action is when the steel beams in a building rust. Oxidation within the body results from the action of free radicals, molecular fragments that contain at least one unpaired or odd electron. Since unpaired electrons are unstable, they attempt to achieve stability by attracting mates: electrons from other atoms or molecules such as cell membranes or the structural components in the interior of the cell. In the process, they injure or destroy these structures. Eventually the free radicals attack the DNA in cells throughout the body. The ensuing chemical damage to DNA is thought to be at least partly responsible for aging. The antioxidants in fruits and vegetables protect against this damage. In the brain the effect on cognitive function is dramatic.

In one study, older persons who ate more than two vegetable servings a day performed as well on cognitive tests as people five years younger. Moreover, this slower rate of cognitive decline held up over a six-year follow-up.

Although vegetables and fruits are often grouped together as cognitively beneficial, the evidence for cognitive enhancement secondary to increased fruit consumption is less convincing. The same study that demonstrated the benefits from vegetables failed to find a similar benefit from fruits. While it's true that vegetables contain more vitamin E than fruits, that probably isn't the main reason vegetables are more protective than fruits. Perhaps some "unknown dietary component that is present in fruit may offset the protective effects of antioxidant nutrients," the authors of the study from the Rush Institute for Healthy Aging concluded.

But before emptying your fridge of fruits, consider that loads of other studies have shown that fruits *are* cognitively protective, especially fruit juices. Both fruit and vegetable juices have been linked to a delayed onset of Alzheimer's disease. It is not necessary to drink large quantities of juices to garner the benefits. Drinking fruit or vegetable juices only three times a week was sufficient in some studies to substantially reduce the risk for Alzheimer's disease.

Why such differences among studies? you may rightfully be wondering. I directed that question to Martha Clare Morris, the principal investigator of the Rush Institute Study.

"This is a very new field and to my knowledge there's really only been a handful of studies that have explored the effect of fruit and vegetable consumption on the human brain," she told me.

Added to the difficulty that we're treading in new territory here, no aspect of diet can ever be considered apart from other lifestyle factors. People who are careful about their diet are also more likely to exercise, watch their weight, not smoke, consume less alcohol, and limit their coffee and tea intake. So which of these factors, or combination of factors, are most important? Researchers on brain health and longevity, in general, can't provide a simple answer to that question because health habits occur in clusters, making it difficult to isolate one habit from another. Nor does everyone react in the same way to a specific substance. For instance, as will be discussed on page 239, new evidence shows that caffeine can be beneficial to brain function. But not everyone can tolerate caffeine. For instance, if I drink more than four or five cups of tea a day (I don't like the taste of coffee), I get too "jived up" to concentrate.

Genetics, too, plays a major role in determining the effects of specific foods on our brain. As a result, conclusions about the various contributions of different foods to good health responses may vary from one population to another. For instance, the vegetable and fruit juice study mentioned above was restricted to Japanese-Americans living in a single county in the state of Washington. Could that fact help explain the different conclusions between that study and the more ethnically diverse one carried out at the Rush Institute?

In order to get answers to questions like that I sought out Carl Cotman, a researcher at the University of California, Irvine, with an international reputation for his acclaimed research on brain-enhancing foods. Cotman knows so much about what we should and shouldn't be eating that during our discussion I found myself running through my mind a scenario in which this professor plans out the menus for all the meals I'll be eating for the rest of my life.

The first thing I liked about Cotman is that he isn't a food ideologue: he suggests that you eat what you want, with the exception of saturated fat–fried foods, cut back on the booze (perhaps with the exception of red wine), and throw away your Davidoff cigars no matter how much you enjoy them (and forget about cigarettes altogether).

While he agrees that genetics and culture play a significant role in determining brain health, he places a lot of emphasis on diet.

"I'm convinced that we know enough to make some highly educated estimates about the foods that are most likely to be beneficial to our brain. I eat a lot more things now containing omega-3 fatty acids, as well as more fruits and vegetables. Fish is especially beneficial, especially if we eat oily fish such as tuna, salmon, herring, mackerel, sardines, sablefish, and trout, which are loaded with the long-chain omega-3 fatty acids.

"In addition, I eat a lot less cholesterol since we found in our studies on rabbits that increasing the cholesterol makes their brain amyloid-loaded. Diet is even more important as we get older, since older cells are less efficient. Until we get more information, the food pyramid is useful and not a bad strategy."

Cotman's fish recommendation has historical precedent. The first hint of the brain-altering properties of oily fish turned up in the 1980s, when psychiatrists noticed that people living in parts of the world such as Japan and Taiwan, where oily fish are part of the regular diet, were sixty times less likely to develop depression than people in countries such as the United States and Germany, where fish is often only a distant second or third menu choice.

In general, the greater the intake of omega-3s in the diet, the lower the incidence of depression. Omega-3s exert an antidepressant effect among not only the clinically depressed but also people who are chronically pessimistic or suffer from frequent bouts of the "blues." Such people often have low levels of omega-3s, and when the levels are boosted as a result of eating more fish, their mood lightens.

But suppose you just don't like fish? Wouldn't taking a

dietary supplement work just as well? Probably not. While supplements such as fish-oil pills are also rich in omega-3s, they don't provide the other benefits of eating fish such as lean high-quality protein low in saturated fat, along with many vitamins and minerals such as selenium. In addition, fish-oil capsules don't have to conform to the same strict FDA regulations as medications. That means that different brands of fish-oil capsules may contain varying amounts of omega-3s.

The best news is that you don't have to eat large quantities of fish in order to reap the benefits. Two servings of fish per week are sufficient. Even if you aren't wildly enthusiastic about fish, two servings (about three ounces per serving) shouldn't be difficult to work into your weekly meal plan. If you have no source of wild-caught fish, you aren't at a disadvantage; in fact, you're ahead of the game. Farmed fish, especially salmon, have more than twice the amount of omega-3s as wild salmon. Scientists speculate that the wild variety is leaner because of the more active lifestyle swimming in the ocean, which burns off some of the fat.

In addition to its effect on mood, omega-3s improve memory and the clarity of one's thinking. We know this from clinical trials, which showed memory benefits in patients in the very early stages of Alzheimer's disease after a switch to a fish diet. It's speculated that the anti-inflammatory effect of the omega-3s interferes with the early stages of the formation of amyloid, the gummy substance that accumulates in the Alzheimer's patient's brain and interferes with the ability of neurons to communicate with each other. But timing was important in the trials: the brain-enhancing effect of omega-3s occurred only during a critical period, two or more years before the establishment of dementia. This corresponds to the same period in which the body's own anti-inflammatory chemicals are elevated in the brain.

Carl Cotman is especially insistent that we do away with long-held distinctions between brain health and heart health. "Whatever enhances the heart enhances the brain. Exercise, nutrition, and stress reduction apply both to brain and heart."

In line with Cotman's "What's good for the heart is good for the brain," three ounces of farmed salmon or six ounces of mackerel a week reduce the risk of death from heart disease by 36 percent. But what about contaminants such as mercury that have been found to be elevated in some oily fish, I asked Cotman.

"Anyone over about twelve years of age who isn't pregnant needn't worry about the sometimes elevated levels of mercury and other contaminants found in some species of oily fish. For the most part, those parts of the brain most affected by toxic contaminants have reached sufficient degrees of maturity that anyone over twelve years of age is unlikely to be harmed." This is also the recommendation of the Institute of Medicine of the National Academy of Sciences: the health benefits of eating fish exceed any risks from contamination by toxic chemicals or heavy metals.

So what is the bottom line here? How best to combine all of these disparate findings to create the closest thing we can to a "brain diet"? To me, after talking to Cotman and others, and conducting my own review of the literature on diet and the brain, I suggest combining into one strategy the various elements that researchers have discovered to be brain enhancing. How to do that without lugging around a textbook on nutrition? By sticking pretty much to the Mediterranean diet, which was originally developed to promote a healthy heart. As Cotman and other experts point out, heart healthy equals brain healthy.

Traditionally, menus in Mediterranean countries are rich in fruits, vegetables, legumes, and cereals. Olive oil and other unsaturated fatty acids are preferred over saturated fatty acids; fish is eaten more often than poultry or meat. Dairy consumption is low to moderate. Alcohol is taken in mild to moderate amounts in the form of wine with meals.

A study of two thousand Manhattan residents averaging seventy-six years of age found that those eating a Mediterranean diet had a 68 percent lower risk of developing Alzheimer's disease. The Mediterranean diet not only reduces the risk of Alzheimer's disease but, according to recent research from Columbia University in New York, may help people already afflicted with the disease live longer.

Drink Moderately and Selectively

The beneficial effects of the regular but moderate consumption of alcohol recommended by the Mediterranean diet came as a surprise to some researchers. Perhaps that's because several generations after the arrival of our Puritan ancestors on these shores, many people in our society still tend to demonize alcohol. (I'm writing these words in Vineyard Haven, Massachusetts, a town on Martha's Vineyard where the sale of alcohol is still illegal.) Certainly, the popular pro-file of the regular alcohol consumer was largely a negative one (increased rates of domestic violence and motor vehicle accidents) prior to the recent discovery of the health benefits of drinking moderate amounts of alcohol in the form of red wine. According to recent research, it's now safe to say that red wine in moderate amounts enhances brain function.

At the moment, nobody has a completely satisfactory explanation of why it is that red wine but not other alcoholic beverages benefits the brain. One theory stresses the impor-tance of resveratrol, a natural compound found in the skins of the grapes used in red wines. Since red wine is fermented with the skins intact, and white wine after the skins have been removed, only red wine contains resveratrol and only red wine exerts a protective effect on brain function—at least that is the popular explanation.

As with many of the studies linking nutrition with brain function, the experimental work on the positive effects of red wine on the brain has been carried out in species other than our own. Giulio Pasinetti, a researcher on nutrition and brain performance at Mount Sinai School of Medicine in New York, described for me one such experiment during a conver-sation after his lecture at a meeting of the Society for Neuroscience.

Over an eleven-month period Pasinetti fed mice in his laboratory an amount of Cabernet Sauvignon equivalent to a daily five-ounce glass for women and two glasses for men. He then tested the ability of the mice to learn to navigate their way within a water maze. Those mice given the wine learned to navigate the maze faster and more accurately than mice drinking just water. In fact, they learned the maze so well that they could remember it when tested forty days later. Coincident with this, the brains of the wine-swigging mice contained decreased amounts of the toxic by-products (amyloid protein) associated with Alzheimer's disease.

While resveratrol is the most likely element in red wine contributing to its beneficial effects on the brain, scientists don't know a lot about its absorption and clearance from the body, its effect on the liver, or even the precise identities of its by-products. Even more important, nobody knows the minimum effective dosage. The resveratrol story contains several other loose ends along with a fatal flaw in its plot structure. Ninety-five percent of resveratrol is destroyed by the digestive system before it enters the circulation. In addition, a 150-pound person would need to drink many bottles (750 to 1,500) of red wine a day in order to achieve the equivalent dose of the imbibing mice— not an advisable scenario. So why not take resveratrol in the form of supplements instead? Before you decide to do that, you should consider the following facts about supplements.

Use Supplements Judiciously

More than half of adults in this country take some form of vitamin or mineral supplement at a total cost of $23 billion a year. The antioxidant market alone grew by 18 percent between 2005 and 2006. Yet there are several reasons for exercising caution with supplements.

First, there is increasing evidence that supplements, in

general, aren't as effective as eating foods containing vitamins and antioxidants in their natural form. Even in nature, only certain forms of the vitamins are biologically active. Of the eight different forms of vitamin E found in nature, only one (alpha-tocopherol) is extracted from the bloodstream— all other forms are excreted. What's more, while powerful antioxidant and other properties can be demonstrated in the test tube, many scientists now question whether similar effects occur in the body. In addition, the natural balance of vitamins and antioxidants found in natural foods is difficult to replicate in the laboratory. Nor is the effect of a supplement necessarily identical to the effect of the natural vitamin. For instance, in a study comparing dietary vitamin E with vitamin E from supplements, only the dietary form reduced the risk of Alzheimer's disease.

Second, some recent studies suggest that the ingredients of some supplements, if taken in excess, may even be harmful. Too much vitamin E, for instance, can thin the blood and increase the risk of bleeding and stroke, especially among people afflicted with high blood pressure. One study from Johns Hopkins revealed that taking 400 IU of vitamin E per day is associated with a higher risk of dying compared with getting vitamin E from food sources.

So far, such findings aren't stopping people from scooping up large quantities of vitamin E, a trend that started in the 1980s. While almost nobody took vitamin E supplements in the 1980s, ten years later an estimated 23 million Americans were downing daily doses. Nor has there been any lessening of this trend.

Multivitamins—currently the most popular approach to

vitamin supplementation—can also prove troublesome if taken in excess. "If one pill is good for you, two or more will be even better" certainly doesn't hold true with multivitamins. Men taking more than one multivitamin a day increase their risk of prostate cancer by 32 percent.

Similar caveats hold for most supplements in healthy people. Unless you are suffering from a deficiency of a certain vitamin, you aren't going to benefit by taking supplements containing that vitamin. But to be fair, all of this assumes that a person is eating an adequate diet—a faulty assumption among older adults.

"With so many people living longer, we're entering into a new stage in our understanding of the older brain," Cotman told me. "The older person's brain needs at least as many vitamins and antioxidants as the brain of a younger person. And yet older people are more likely to be lacking them because their diet is often inadequate as a result of their tendency to skip meals and eat less in general. The older brain needs supplements because they help prolong life and extend the years of healthy living. Smokers and people who drink a bit more than they should are also likely to be deficient in omega-3 fatty acids and other needed vitamins and antioxidants. The older person eating an insufficient or faulty diet can derive a lot of benefit from supplements."

So what practical conclusions can one draw about diet and the brain? Here's a summary of the most commonly accepted dietary formula for maintaining optimum brain health: maintain low weight, cut way down on saturated fats, eat fish and other foods containing high amounts of omega-3 fatty acids, decrease your intake of red meats, take in more fruits and vegetables, drink red wine in moderation, and sensibly supplement your diet with vitamins and antioxidants.

Exercise More

While nutrition is probably number one in importance as a stimulator to brain development, exercise runs a close second. That wasn't always true. I grew up as a member of a generation that made firm distinctions between mental and physical exercise. During my college and medical school education "jocks" were routinely made fun of; the smartest people (or those who earned the highest grades in school) usually did little exercise, rarely tried out for sports, and by today's standards led physically inactive lives. Now all of that has changed for the good, thanks to the success of such scholar-athletes as New York Knicks NBA Hall of Fame winner and former U.S. senator Bill Bradley and Sebastian Coe, long-distance runner and British MP—among others. We now know that an optimally functioning brain goes hand in hand with general good health. And since the brain is a biological organ governed by the same factors as all other body organs (a need for oxygen, glucose, and other nutrients, along with adequate blood perfusion), physical exercise, which increases all of these things, plays a leading role in enhancing brain function.

According to recent findings, consistent aerobic exercise leads to the growth within the brain of new capillaries—the tiny vessels bathing the neurons with nutrients. Accompanying this is an increase in the number of interconnections between neurons and the number of brain cells in the memory-encoding hippocampus. These enhancements result from exercise-induced elevations in the levels of nerve growth factors such as brain-derived neurotrophic factor (BDNF) and other nutritive tonics. Since a physical workout increases all of these factors, the end result is a better connected, more adaptive brain capable of thinking better and faster, processing information more efficiently, and generally working better overall.

According to Arthur Kramer, working out of the Beckman Institute at the University of Illinois, older adults with a lifelong history of cardiovascular exercise have better-preserved brains than sedentary people of the same age. The structural preservation is greatest in the frontal, temporal, and parietal lobes—three of the most important areas of the brain for mental processing.

Activation of the prefrontal area focuses our attention inward on what we're *thinking;* temporal and parietal activation focuses attention outward on what we're *doing.* Because these areas are better preserved among exercisers, longtime exercise adherents of any age perform better than couch potatoes when it comes to memory and the ability to remain focused.

"Although physical fitness training broadly influences a variety of mental processes, the largest gains involve the executive control processes," according to Kramer. "Interestingly, these are the processes that show substantial age-related decline." These are also the processes that differentiate older from younger workers (as will be discussed in more detail in the chapter on compensating for age-associated brain changes). Both mental and physical exercises can narrow the age-related performance gap.

The most interesting aspect of Kramer's research findings was the short time frame required to bring about these changes. "Only six months of regular aerobic exercise increases brain volume," Kramer, a bearded and convivial middle-aged man, told me during a conversation following one of his well-attended lectures. "We have here a simple and inexpensive mechanism to roll back some of the normal age-related losses in brain structure. We now know that brain volume loss is not an inevitable effect of advancing age. Relatively minor changes in activity level can go a long way in offsetting and minimizing brain aging."

Anyone of any age who walks three times a week for forty-five minutes will reap the following benefits: sustained levels of cerebral blood flow; an improvement in focused attention; increases in gray matter volume in regions of the frontal and temporal lobes; and restoration of some of the losses in brain volume associated with normal aging. In anyone over sixty, the benefits are even greater. A daily one-mile walk will reduce the likelihood of dementia by 50 percent.

Exercise also brings about positive changes in the brain's chemical messenger system. For instance, an antidepressant drug that affects three of the brain's main neurotransmitters (norepinephrine, epinephrine, and serotonin) works even better when combined with exercise. As a result, neuro-psychiatrists like myself are now encouraging our depressed patients not only to take their medicines, but to become more physically active as well. Since exercise and antidepressant medication affect some of the same neurotransmitters (depending on which antidepressant is taken), an increase in exercise may allow for a reduction in dosage of the drug.

One final point that should prove even more persuasive to anyone who dillydallies about exercise: Exercising at least three times per week lessens the odds of coming down with Alzheimer's disease. Specifically, you can decrease the likelihood from about twenty in a thousand to thirteen in a thousand. Further, the earlier in life you begin exercising, the faster the processing speed of your brain when you're sixty-five years of age or older. This doesn't mean that if you were physically active earlier in life that you don't have to continue to exercise. But it helps explain why some middle-aged adults who aren't currently physically active but who exercised when they were children or young adults remain mentally sharp in their later years. The question is: Why did they give up the exercise habit?

Although there isn't a simple answer to that question, one contributor, in my opinion, is the basically boring nature of most exercise programs. In fact, the very term "exercise program" turns off many people (me included).

As an alternative to a "program," consider working out using the Nintendo Wii video-game system.

The original Nintendo system with its thumb-driven controllers was designed for entertainment and provided no exercise opportunity for any part of the body other than the hand. But the new Nintendo Wii system (pronounced "we") uses handheld motion controllers that require the player to employ the same arm and body motions used in real-life tennis, golf, bowling, baseball, and boxing. The controllers are precise enough to track and mimic the body movements employed by real players engaged in these different sports.

In Nintendo Wii tennis, for instance, you can return a shot via a lazy lob, but as in the real game, you'll do a lot better if you hit the ball forcefully and precisely in response to the simulated tennis situation on the screen. Other games like bowling offer you the chance to compete against other participants. But whatever game you choose, the exercise benefits are much the same as those from real bowling, minus the subsequent aches and pains that are typically endured the day after holding and rolling a fourteen-to-sixteen-pound bowling ball.

WiiFit, a fitness-themed software program, is even more physically stimulating. For this program players stand on an electronic platform containing sensors that translate the pressure created by body motion into televised motion on the monitor. In a ski-jump program, for instance, the Wii skier has to squat down and spring up at the precise second that he reaches the end of the ramp. Precision and balance combined with physical endurance are demanded in the boxing program as players simultaneously weave, duck, and throw combination punches. Thanks to such a program, you can pretend you're Oscar De La Hoya while sparing yourself the brain-jarring head punches associated with real boxing.

If you're not drawn to such games, find the prospect of spending time in a health club less than thrilling, and don't want to devote inordinate amounts of time and energy to exercise, you might want to try my personal exercise prescription:

At least three times a week I follow Arthur Kramer's

advice and walk for thirty to forty-five minutes at a brisk pace in different locations around Washington. By walking in varied neighborhoods I can combine exercise with learning more about the city. As I walk I remain fully alert to my surroundings; when I encounter something that interests me, I stop. This combination of physical with mental activity keeps me motivated in a way that a trip to a health club doesn't. (To be fair, a health club offers opportunities for meeting other people and thus can serve as an antidote for loneliness, which is linked to Alzheimer's disease.) But if loneliness or isolation isn't a problem, you may find walking (perhaps combined with some jogging) to be your best means of getting the health benefits of exercise without being tethered to a repetitive, joyless routine.

Think of exercise according to a continuum model. At one end are those who claim that exercise is now an optional leisure activity, since it's no longer necessary for physical survival (no more wild animals to elude). Not surprisingly, those who hold such views tend to be nonexercisers. At the other end of the spectrum are those intensely devoted exercisers who script their entire lives around exercise. Long-distance running is especially appealing to the latter group.

Although I'm somewhere in the middle of the exercise continuum, I freely admit that evidence exists favoring the view that maximal brain benefits can come only from prolonged and highly effortful exercise such as long-distance marathons or triathlons. It's correctly pointed out that such challenges increase not only cardiovascular fitness, but also strengthen cognitive and psychological traits like focus and endurance.

The Japanese novelist Haruki Murakami captures this link between extreme physical effort and mental endurance: "You've got to have physical strength and endurance to be able to spend a year writing a novel and then another year rewriting it again ten or fifteen times. Stamina and concentration are two sides of the same coin. I sit at my desk and write every day, no matter what, whether I like it or not, whether it's painful or enjoyable. I do this day after day, and eventually—it's the same as running—I get to that spot where I know it's been what I've been looking for all along. You need physical strength for something like that. . . . It's like passing through a wall. You just slip through."

Some experts claim that for walking or running exercises to be beneficial they must involve long distances. Certainly from the evolutionary point of view, that might make sense: although humans are slow runners compared with other animal species, we excel at endurance running and can outrun most other animals over long distances. But does this necessarily suggest that we will gain the most brain benefits by long-distance running? Since at this point nobody can answer that question for certain, the choice is an individual one. (As mentioned earlier, I'm persuaded by Kramer's findings of the benefits of a forty-five-minute walk three times a week.)

So should you follow a similar take-it-to-the-limit philosophy of exercise, or are the forty-five-minute walks three times a week sufficient? On this point, you will have to decide for yourself the level of exercise intensity you wish to pursue. One word of caution: If you've been physically inactive most of your life and now wish to engage in highly strenuous physical exercise, you should first undergo a physical exam and cardiac stress test, then decide about your goals. Murakami took up exercise primarily to improve his mental stamina. You may desire simply to lose weight, increase your energy, and improve your overall health. To accomplish this you don't have to participate in triathlons. But whatever your motivation, you can be confident that exercise of any intensity will improve your overall brain health.

Pay Attention to Sleep and Naps

After nutrition and exercise comes the third basic component of the brain health troika: sleep.

In our hard-driving 24/7 culture, sleep gets little respect. Many of us consider it a useless though unavoidable "downtime" that cuts into our productivity. That sentiment at least partially explains why, on the average, people are sleeping forty-five minutes less today than their ancestors only twenty-five years ago. Yet sleep isn't something that we can simply do away with.

I remember that the hardest part of my medical school and internship training was the need to balance high performance with insufficient sleep. After a day spent in a busy clinic, I worked for much of that evening and into the early hours of the next morning in the emergency room. After perhaps two hours of sleep, from four a.m. to six a.m., I had to be ready to put in another full day on the wards working with patients. Only at the end of that stint was I free to "crash" until the next morning, when I started another thirty-six-hour sleep-deprived endurance contest. I continued that every-other-night schedule for a whole year during my internship.

Fortunately, working despite severe sleep deprivation is now a thing of the past for doctors-in-training. Today they work much more sensible hours, and for good reason. If we're sufficiently sleep deprived, our judgment and performance are seriously affected; if we're sleepy enough, for instance, our driving skills approach those of individuals observed with alcoholic intoxication.

At the other extreme, the more we sleep (up to a point, of course) the better we perform: high-achieving students usually sleep more than their less successful counterparts, according to studies correlating sleep and student grades. But how much sleep do we actually need for our brain to function at its best? And why does our cognitive performance deteriorate so dramatically when we don't get enough of it?

I posed these questions to sleep researcher Clifford Saper of Harvard University. "Think of the brain in terms of the lowly tomato," he told me. "Tomatoes grow in the summer by carrying out photosynthesis during the day and storing the energy derived from that process. At night they use that stored energy for growth. A similar process occurs in the brain. During the day it takes in a tremendous amount of energy in the form of information. In order to store that information the brain has to make long-term changes in its structure at the level of synapses. To do that, it uses the same cell-to-cell signaling systems that it used during the day to store the information in the first place. At night the brain, like the tomato, uses its stored energy for growth, making the necessary structural changes for information storage."

Thus, contrary to popular opinion, the brain doesn't "turn off" during sleep. The sleeping brain has distinct activity patterns and electrical signatures consisting of the periodic repetition of four basic phases. Rapid eye movement sleep (REM, also known as dream sleep, since this is the phase when dreams occur) is characterized by brain waves that most resemble those found during wakefulness. The other three phases, collectively referred to as non-REM (NREM) sleep, display electrically distinctive substages of sleep that regularly alternate with one another about every ninety minutes during the night.

During REM sleep the brain makes those structural changes mentioned by Saper. Further, during REM the brain is reusing some of the same synapses used while awake. This explains why most of our dreams relate to events, situations, and concerns that happened during the previous day.

A sleep experiment carried out several years ago provides a neat proof of the association of daytime events and nighttime dreams. It involved people who had spent many hours during the previous day playing Tetris. That night they dreamed of falling shapes—suggesting that the brain synapses that had been used for Tetris were more easily activated during sleep, hence the dreams of falling shapes. Other synapses that hadn't been involved in Tetris remained deactivated during sleep.

A similar process takes place during learning and the establishment of memories. During sleep the brain "replays" the same pattern of activity that occurred when something was learned during the previous day. Thus a PET scan "snapshot" of the brain's activation patterns taken during daytime learning is virtually identical to the PET "snapshot" of activity later that night. In one famous experiment illustrating this, Pierre Macquet of the University of Liège, Belgium, monitored brain activity in men playing a virtual reality game in which they learned to wend their way through a virtual town. The same regions of the brain (the hippocampus) that were activated when the men visually explored the virtual town also activated that night during slow wave sleep.

Furthermore, the more a person learns during the day, the greater the amount of replay during the night, according to a study of computer video game players. At night, when the game players were asleep, the same memory encoding vigorously came online—the identical finding in the Macquet study. "How strongly the hippocampus came back online at night predicted how much better the players would be the next day," says Matthew Walker, director of the sleep and imaging laboratory at Beth Deaconess Medical Center in Boston. "The more the brain learns, the more it demands from sleep at night." Walker's comments provide a partial explanation for the finding, mentioned earlier, that high-achieving students require more sleep than their less studious counterparts.

For learning to take place, a certain amount of time must pass in order for "consolidation" to take place. "Consolidation goes beyond simply stabilizing or fixating memories— it enhances them as well," according to Walker. In addition, the timing of these two processes differs.

Consolidation occurs whether we're awake or asleep. Enhancement, in contrast, occurs primarily if not exclusively during sleep. "This 'off-line' effect during sleep can produce additional learning," says Walker.

Walker's research suggests a strategy for taking advantage of the distinction between consolidation and enhancement: since sleep-induced enhancement involves improving upon what you have learned while awake, you should schedule periods of sleep between practice sessions. That way you benefit from the fact that sleep restores, refreshes, and enhances the local brain circuits mentioned by Saper that are involved in skill learning.

To get a feeling for how all this works, think back to when you learned a new tennis stroke. Initially your performance improved as a result of repeated repetition. But if you continued beyond a certain point, additional practice led to deterioration in your performance (the "overpractice" effect known to every athlete). It's speculated that the performance deterioration reflects a selective fatigue of those brain regions used during learning that new stroke.

What is the proper response to that predictable sequence of improvement followed by a falloff in performance? Stop practicing at the first sign of deterioration in your performance and don't return to tennis practice until after a night's sleep. Thanks to the consolidation-enhancement nexus, you'll wake the next morning with the relevant brain circuits refreshed. As a result, your performance will incorporate all of the things you learned during your practice session the day before.

In general, you can expect a selective enhancement for your weakest areas of tennis or golf performance if you get in a night's sleep before taking your next lesson. This holds true for any activity requiring the learning and repetition of skilled muscular movements. For reasons that aren't entirely clear, sleep serves to improve one's overall performance by selectively enhancing those areas that are most in need of improvement.

Getting a good night's sleep increases efficiency and helps you perform at your best. In one experiment confirming this, a night's sleep restored accuracy and speed to the peak level achieved during the initial learning phase in a simple typing task.

"A key role of sleep in learning and memory is to restore the local brain circuits participating in the learning of a skill, thereby maximizing the benefits of presleep training," Walker told me. In short, if you learn something while awake, you can increase your chances of remembering it by "sleeping on it."

Here's one more important detail about consolidation and memory: The initial period of memory consolidation occurs within the first six hours after learning. In practical terms, if you try to learn a second unrelated skill during that period, you will interfere with what you just learned during your initial effort. This is especially true in physical activities (motor programs, as neuroscientists refer to them). So after the completion of a tennis lesson you should not follow that up during those crucial six hours with instruction in golf or any other motor skill lest learning that second activity interfere with your memory consolidation for the tennis. Researchers discovered this principle from an experiment during which subjects learned a series of finger movements followed by an immediate follow-up session in which the participants learned the reverse sequence. When taken in succession, learning that second sequence impaired the performance for the first sequence. But the two sequences were equally well retained if several hours of sleep separated learning the two skills.

So if you want a selective improvement in your weakest areas of tennis or golf, remember the six-hour consolidation rule and make certain you get in a night's sleep between lessons. This holds true as well for any activity requiring the learning and repetition of skilled muscular movements. Although no one knows exactly why, sleep improves one's overall performance by selectively enhancing those areas that are most in need of improvement.

Work to Resolve Insomnia

If your sleep is disrupted, your recall for what you've learned declines, your memory returns to a labile, easily modified state, and consolidation doesn't take place. Sleep researcher Robert Strecker of Harvard University described for me his experiment that led to this conclusion.

Strecker placed rats in a water maze—the equivalent of a rat swimming pool filled with murky water concealing a small platform hidden beneath the surface. Since rats aren't long-distance swimmers, they need to find the platform as quickly as possible and climb up on it. Once having learned the location of the platform, rats tend to remember it and swim quickly to the platform when placed back in the water. Sleep-deprived rats that had been purposely woken up periodically during the night before the water maze test took longer than well-rested rats to find the platform.

"This suggests that sleep fragmentation resulting from periodic awakenings affects the hippocampus, the area responsible for storing spatial maps in the brain," Stecker says.

At the molecular level, sleep deprivation leads to the accumulation of stress hormones in the hippocampus, which in turn stunts the growth of cells that establish new memories. Moreover, this effect of sleep fragmentation is difficult to reverse. Even after being allowed to sleep undisturbed the following night, the rat's memory for the platform location remained impaired.

Except in nightmares, sleep fragmentation in our own species doesn't involve desperate searches for platforms hidden beneath murky water. Rather, our sleep fragmentation involves bouts of insomnia, defined as one or more of the following symptoms for at least a week: taking more than an hour to fall asleep; wakefulness for over half an hour during the night; failure to feel refreshed and rested upon awakening. This combination of symptoms is common. Sleep specialists estimate that 70 million Americans suffer from insomnia.

Sleep specialists have confirmed what all of us have experienced at one time or another: Sleep deprivation wreaks havoc on our mental and physical health. The risks of heart attacks, stroke, and mental breakdowns (especially among people with bipolar disorder) all rise when people are deprived of adequate sleep.

Yet sleep deprivation is a common problem even among

"normal" people. Advertisements for prescription sleep aids (43 million prescriptions written per year) make up a goodly proportion of television commercials. With 26 million people working as shift workers (20 percent of whom simply can't tolerate such a work schedule and eventually give up their jobs), the disruptive effect on the brain's normal rhythms is a nationwide challenge.

Thanks to PET scans we now know what is going on in the brain of insomniacs. Even after finally getting to sleep the insomniac brain shows heightened activity in the arousal circuits. And since this activity prevents deep restorative sleep, insomniacs experience daytime problems in learning, memory, concentration, and mood. Indeed, psychiatrists are now speculating that insomnia, a frequent accompaniment of depression, may be the *cause* rather than the result of depression. In support of this theory, depressed people recover more quickly if, as part of their treatment, they are given sleeping pills along with an antidepressant instead of only an antidepressant.

Even among insomniacs who are not clinically depressed, sleeplessness leads to fretfulness and worry about things that next morning are recognized as not worth fretting over. Think back to the last time you had trouble getting to sleep. Odds are you weren't lying there in a relaxed frame of mind while thinking how wonderfully everything was going in your life. More likely you were running one or another personal doomsday scenario through your mind.

"People are not awake because they are worrying, they're worrying because they're awake," according to Michael L. Perlis, director of the sleep and neurophysiology laboratory at the University of Rochester. In other words, insomnia creates a worried, anxious frame of mind rather than serving as one of its symptoms. In such a state your brain is unable to restore and refresh the local brain circuits involved in memory.

Insomnia is such a debilitating problem because disturbed sleep, with the accompanying daytime sleepiness, is incompatible with optimal brain functioning. So if you suffer from insomnia, take the necessary steps to resolve it. Start by seeking professional help to find out if you suffer from some of the easily treatable causes of sleep disruption such as obstructive sleep apnea. Devices exist that can normalize breathing patterns and restore restful, refreshing sleep. If after investigation a specific cause isn't found, you are likely to benefit from occasional use of sleeping pills. Medications are on the market now that target different neurotransmitters while exerting minimal effects on the normal brain wave patterns associated with normal sleep.

Embrace the Power of Naps

After you've resolved your nighttime sleep problems you can use naps as an additional brain enhancer. A daytime "power nap" will produce nearly as much skill-memory enhancement as a whole night of sleep, according to the findings of Mathew P. Walker.

In an experiment testing finger dexterity, Walker taught student volunteers skilled finger movements similar to piano scales. After learning the movements half of the students took a sixty-to-ninety-minute nap while the others remained awake and continued with their day. When retested later that same afternoon, those who napped did 16 percent better than those who did not nap.

"A daytime power nap produces nearly as much off-line memory enhancement as a whole night of sleep," Walker told me. In order to do this, the brain selectively increases brief bursts of electrical activity called "sleep spindles." Walker believes that these spindles trigger chemical reactions within brain cells that "instruct" specific brain circuits to strengthen connections, and thereby enhance memory.

"The brain selectively increases spindle activity in local brain circuits, thereby discretely targeting those regions in the brain that have recently formed new memories. Sleep spindles appear to make a selective and critical contribution to improving our motor memories at night and across power naps during the day," said Walker.

Fortunately, the benefit of naps isn't confined to learning motor skills, because most of the things we want to remember involve facts, words, and concepts rather than motor skills.

In an experiment confirming this, Matthew Tucker at the City University of New York asked volunteers to memorize pairs of words. They were then tested immediately afterward and a second time six hours later. Those who had been allowed a nap of less than one hour (far too short a time for REM sleep to occur in the normal person) scored 15 percent better than the nonnappers.

Naps play an especially critical role in the lives of highly creative people. When performance psychologist K. Anders

Ericsson examined the diaries of expert musicians, he found that most of them could engage in concentrated focused practice for only around an hour. After that, their concentration began to falter and their performance declined. To compensate for this turndown in performance, they spent more time napping. These recuperative naps restored their ability to maintain the high levels of concentration required for their creative efforts.

At this point you're probably wondering: But how do I force myself to fall asleep for a nap less than an hour long? And doesn't taking a nap make it more likely that I'll not be able to get to sleep later that night?

Sleep can't be forced. It has to flow naturally from a state of relaxation and surrender. Indeed, sleep involves a paradox: The more effort you expend trying to fall asleep, the more awake you become. Instead of trying to force yourself to sleep, you should set an alarm for thirty minutes (or ask someone to awaken you after that time) and lie down with the intention of simply relaxing. After a few days of doing this, you'll find yourself drifting off into a brief but refreshing sleep. I discovered this several years ago when I was teaching an evening course on the brain at the Smithsonian Institution in Washington, D.C.

My lecture, and the questions-and-answers following, went on from six to eight p.m. As I discovered, that schedule proved exhausting, especially on busy days when I was already beginning to feel fatigued after a day in which I wrote from seven to nine a.m., then treated a steady stream of patients from ten until four. In order to recharge my energy before my lecture, I decided to try simply turning off the lights in my office at four-thirty and resting on the couch.

The first few attempts at napping were failures. I didn't fall asleep and didn't feel any less tired when I got up. But I kept at it and after a few days of simply "relaxing" and not trying to "make myself" fall asleep, I started drifting into a light sleep within minutes of lying down. Now I can fall into a light sleep for fifteen or twenty minutes almost every time I lie down. Most important, I can deliver evening lectures without feeling fatigued and still remain fully engaged afterward with questions from the audience.

The second question—Do daytime naps interfere with sleep later that night?—requires a more nuanced answer. Yes, if a nap goes on too long it does interfere with sleep quality and duration later that night. That's why I suggested resolving any nighttime sleep difficulties before establishing the nap habit. In general, a nap lasting more than an hour is probably too long. The key is to find your own personal cutoff point separating a nap that increases your mental sharpness from one that makes you feel groggy upon awakening and, later that night, results in insomnia.

In addition, if a nap goes on for too long, cognitive performance immediately after the nap is worsened rather than improved. Your personal optimum nap length is best estimated by gauging how you feel after the nap. If you feel mentally refreshed and are eager to get back to a mentally challenging task after your nap, you are "in the zone" for achieving your goal of using naps to enhance your cognitive performance.

Just as important as getting enough sleep is thinking about sleep in the right way. Stop thinking of sleep and naps as "downtime" or as a "waste of time." Think of them as opportunities for memory consolidation and enhancing the brain circuits that help skill learning. Nor should you feel guilty about sleep. It's just as crucial a part of successful brain work as the actual task itself.

PART THREE

Specific Steps for Enhancing Your Brain's Performance

Now that we've covered the basics of diet, exercise, and sleep, let's explore the specific steps you can take to achieve a super-power brain. Some of the world's most prestigious brain experts have suggested to me that anyone who wishes to develop and maintain an optimally functioning brain must work on the following functions:

Attention: This must be rock solid. If you can't maintain focus or concentration, you can't marshal the effort needed to improve performance in the other brain functions. Attention in the mental sphere is equivalent to physical endurance in the physical sphere.

The Japanese novelist Haruki Murakami—a former couch potato turned marathon runner—captures this link between physical effort and mental concentration, as we have already seen: "Stamina and concentration are two sides of the same coin. I sit at my desk and write every day, no matter whether I like it or not, whether it's painful or enjoyable. I do this day after day . . . it's the same as running. You need physical strength for something like that."

Memory is a natural extension of attention. If you attend to something you increase your chances of remembering it. Memory also anchors us; in many ways we are the sum total of our memories. We learn from past experiences—but only those that we can remember. As we lose memory, we lose parts of our experience and can suffer a form of identity disorder. But we can combat this by strengthening our memory, by reaching into the past and recapturing earlier thoughts, experiences, and emotions. In the process we unify our personality by integrating our past with our present and our imagined future. In essence, we create who we are.

Traditionally, memory can be broken down into three broad types: sensory memory, long-term memory, and short-term or working memory.

Sensory memory consists of the brain's initial recording of physical sensations as they impinge on our sense organs. Iconic memory (things that we see) and echoic memory (things that we hear) are the main forms of sensory memory. Typically sensory memory occurs outside our awareness, but there are ways that we can make the process conscious and thereby improve it.

Long-term memory refers to information that becomes a permanent part of us (information about our work, our relatives and friends; basic facts about our culture such as holidays, movies, and television shows; income tax deadlines). Long-term memory can be strengthened by practice. You can store as much information as you want in long-term memory over your lifetime without ever exceeding its capacity. Indeed, long-term memory is essentially infinite. Vocabulary is the best example. We can learn new words and phrases throughout our lives, no matter how old we become.

Just because long-term memory is potentially infinite doesn't mean that you can recall everything that you've ever learned. Conversely, forgetting a piece of information doesn't mean that it has disappeared forever; rather, for one reason or another you cannot retrieve it. Thus people with "good" memories don't necessarily store more information in their long-term memory—they're just better at accessing it. As a general rule, memories are best retained and retrieved when they are linked with an image or an emotion.

Working memory: Also known as short-term memory, working memory involves the most important mental operation carried out by the human brain: storing information briefly and manipulating it. You're using your working memory whenever you balance two or more competing thoughts, shifting at will from the first thought to the second thought while holding the original thought in a kind of suspended animation. The process is similar to shifting documents on your computer by toggling from one to another.

Working memory differs from long-term memory in an important way. While long-term memory is for the long haul—establishing memories that become permanent and available for future retrieval—working memory is for "right now." For instance, it was helpful three nights ago for you to temporarily store in your working memory what your waiter looked like so that you could distinguish him from the other waiters in the crowded restaurant when you wished to get his attention. Today you have no need to remember him and, as a result, he is no longer a part of your working memory and occupies an infinitesimal portion of your long-term memory—if you visit that restaurant again you may or may not recognize him.

Think of working memory as the conduit for long-term memory: most things encoded in long-term memory are first processed in working memory. It usually requires concentrated effort by your working memory to encode all of the information now residing in your long-term memory. As a child you learned the alphabet and multiplication tables by laboriously memorizing them, i.e., running them through working memory. Eventually you could recite the alphabet and do the sums automatically as a result of shifting the alphabet and the multiplication tables from working memory into long-term memory. From that point, if you were asked to spell a simple word or carry out a multiplication contained in a table (How much is 7 times 7?), the answer sprang to your lips from long-term memory without the need to first process it using working memory.

Working memory can be compared to an air traffic controller simultaneously keeping track of incoming and departing flights. The controller must be capable of shifting her attention from plane to plane while simultaneously keeping "in mind" (i.e., in working memory) each plane's coordinates. Most important of all, working-memory capacity is correlated with intelligence, especially performance on intelligence tests. Some psychologists argue that general intelligence consists of sharply honed working-memory skills.

According to Henry L. Roediger, the distinguished chair of the Department of Psychology at Washington University in St. Louis, "The more capacity people have to hold information in mind while they think, the more intelligent they are."

Mental exercises: We hear a lot these days about mental exercise. But what exercises are most important to carry out? For mental exercises to be beneficial to brain health, they should be tailored, I'm convinced, to each person's interests and proclivities. In other words, the benefits accrue only if you're doing something you enjoy. At the moment, I'm sitting at my writing desk, which overlooks woods across the street. I'm staring at the computer screen while writing the sentence that you are now reading. Meanwhile, my wife sits one floor below me in the company of her ever patient dog and my irascible parrot while finishing the *New York Times* crossword (she's already finished the *Washington Post* crossword, along with the sudoku for the day). In a few minutes she will get up to go to her office. Question: Which of us is engaged in the greater brain challenge?

My brain stimulation comes from writing books or book reviews. For my wife, sudoku or crossword puzzles work just as well. So there isn't any "best" way to enhance brain function that can be applied to everyone. Instead, different approaches are appealing to different people.

Another important point: Mental exercise differs from physical exercise in that it provides specific limited benefits, while physical exercise bestows generalized benefits. If you carry out a fitness program that's well designed for your body, your general health will improve: you will lose weight, lower your blood pressure, and regulate your blood sugar level. Mental exercises, in contrast, tend to benefit specific mental functions. If you engage in memory exercises, for instance, you'll improve your memory but won't do much to increase your logical powers. Nor will increasing your facility with numbers do much to improve your fine motor skills. Because of this tendency for mental exercises to provide limited and circumscribed benefits, a program for enhancing brain performance must include efforts aimed at improving the functions mentioned above along with seven other key brain functions:

Visual observation
Fine motor skills
Tactile perception
Logic
Numbers
Imagination
Visual-spatial thinking

Before exploring how to enhance these key brain functions, let's say a few words about research that is challenging the traditional view that our brain's power and efficiency are determined largely by our genetics.

The Intelligence Conundrum

Intelligence is the most striking example of this revolutionary transformation in our thinking. New findings have brought experts to believe that we can increase our intelligence via our own deliberate efforts. This is a revolutionary change in attitude. In only a few years we've progressed from the belief that "by all means study hard but no matter how hard you work, don't think you can exceed the intellectual performance of those who are lucky enough to have been born more intelligent than you" to the more exciting claim that—with the probable exceptions of the Mozarts and the Shakespeares of the world—the principal limitation on intelligence is how hard a person is willing to work. (No doubt this was what Thomas Edison had in mind when he defined genius as "one percent inspiration and ninety-nine percent perspiration.")

What is intelligence? According to one not very helpful definition, "intelligence is what intelligence tests measure." But defining what you are trying to measure by referring to the readings of the measuring instrument is a flawed approach. Still, estimating a person's intelligence on the basis of his performance on IQ tests remains firmly entrenched in both the popular and professional imagination.

After a lifetime of test taking in a culture obsessed with standardized testing, we've been conditioned to believe that (1) intelligence is best measured by IQ tests; (2) IQ is genetically determined, with the environment playing only a secondary role; and (3) individual IQ gains over an individual's life span tend to be modest. None of these assumptions is correct, according to intelligence researcher James R. Flynn.

Flynn is world-famous for his discovery of an impressive increase in IQ scores over the past hundred years in the world's industrialized countries (the so-called Flynn effect). Since genetic changes aren't likely over such a comparatively short time (the early 1900s to the present), genes alone couldn't be responsible for the IQ gains. The direct effect of genes on IQ accounts for only 36 percent of IQ variance, with environmental differences making up the remaining 64 percent, according to Flynn.

When I first learned of Flynn's claims, I must admit I had my doubts. For one thing, I've encountered people during my educational career who consistently outperformed all of their classmates (including me), despite their seeming to put in minimal effort. But in at least one case—a classmate in medical school—a bit of deception was involved. While the rest of us engaged in marathon study sessions, he spent his time playing table tennis or going to movies—or so he said. The truth came out when his girlfriend inadvertently mentioned that he spent hours studying at her apartment. When I asked him why he was studying as much as the rest of us but took such efforts to pretend otherwise, he replied that he did it because he thought it was "cool." In illustration of his point he showed me a cartoon of a mother duck leading her three ducklings across a pond. At the halfway point she turns to them and says, "Make it look easy, but underneath paddle like hell." In short, my classmate wasn't an example of the preeminent influence of genes; although he pretended otherwise, he wasn't any smarter than the rest of us. Environment (time spent studying) played a large role in his scholastic performance.

Twin studies provide more difficult-to-discount evidence favoring genetics rather than environmental influences like individual effort (i.e., hard work) as the basis for intelligence. In general, identical twins separated at birth and raised apart end up with very similar IQs. This would *seem* to prove that genes exert the most powerful influence on intelligence. But not necessarily. To understand why, Flynn draws on an analogy from basketball.

If on the basis of their genetic inheritance both members of a separated-twin pair are tall, quick, and athletically inclined, they are both likely to be attracted to basketball, practice assiduously, play better, and eventually attract the attention of basketball coaches capable of transforming them into world-class competitors. Other twin pairs, in contrast, endowed with shared genes that predispose them to be shorter and stodgier than average will display little aptitude or enthusiasm for playing basketball, and end up as spectators rather than players.

Flynn suggests a similar environmental influence on genetic inheritance in regard to IQ: twins with even a slight genetic IQ advantage are more likely to be drawn toward learning, perform better in school, and be admitted to the most competitive universities. In the process their IQ levels are likely to increase even more.

"There is a strong tendency for a genetic advantage or disadvantage to get more and more matched to a corresponding environment," Flynn says. As a result, the environment will always be the principle determinant of whether or not a particular genetic predisposition gets fully expressed. This holds true not just for IQ, but also for other cognitive processes such as memory and mental acuity. In practical terms, this means that our cognitive powers can be enhanced through our own deliberate efforts.

Can a person increase his or her intelligence? The answer involves the difficulty, mentioned above, of coming up with a universally acceptable definition of intelligence. But rather than allowing ourselves to become entangled in knotty and ultimately fruitless attempts to define intelligence, let's concentrate instead on identifying *some of the traits commonly recognized to be associated with heightened intelligence and enhanced mental abilities*.

- *Mental acuity (fluid intelligence)*. Fluid intelligence comes into play whenever we solve a problem by employing a free-form approach rather than relying on previously acquired knowledge. For instance, if someone with no medical training is forced under emergency circumstances to tend to a sick person, the amateur caregiver must rely on a seat-of-the-pants approach to compensate for his or her lack of medical knowledge. This fluid intelligence is processed by the prefrontal lobes, one on each side of the brain. Damage to these areas exerts a devastating effect on fluid intelligence and results in severe impairments in problem solving. The person with frontal lobe damage can lose this ability to come up with improvisational approaches, less than perfect but best under the circumstances, to problems and challenges.

- *Knowledge and information (crystallized intelligence)*. In contrast to fluid intelligence, crystallized intelligence relies on previously acquired knowledge. When a trained physician diagnoses and treats an illness, she's relying on crystallized intelligence: the practical application of her years of medical training. In contrast to the person with no medical training, she doesn't take a tentative unstructured approach to her patient's illness; her store of knowledge and information allows her to respond quickly and decisively. The greater that store of knowledge, the greater her control over the treatment situation. One more distinction from fluid intelligence: crystallized intelligence isn't as affected by brain damage either in the prefrontal lobes or elsewhere.

- *Memory*. Learning new information isn't helpful unless it can be recalled later. Anything that increases one's memory powers increases access to everything learned.

- *Curiosity.* The more curious we are, the more we learn.
- *Speed of information processing.* In general, faster mental speed is associated with higher intelligence. Typically we describe someone who isn't very intelligent as "slow"—an intuitive appreciation of what experimental psychologists have found in reaction time tests. In one test measuring speed of information processing called the odd-button-out test, the subject rests a finger on a "home button" while looking at a screen showing three target buttons. If two of the target buttons illuminate, the subject moves his finger from the home button and presses the third button on the screen as quickly as possible. If only one target button lights up, the subject presses that button. People who respond the quickest in the odd-button-out test have an above-average IQ.
- *Ability to think in abstract terms unrelated to specific applications.* The increase in IQ in industrial countries over the last hundred years is the result of a progression from concrete to abstract levels of thinking and understanding (i.e., a chair and a table are similar not because they both have four legs—a concrete response—but because they are both items of furniture).

In addition, Flynn suggests that when speaking about intelligence, we "dissect it into solving mathematical problems, interpreting the great works of literature, finding on-the-spot solutions, assimilating the scientific worldview, critical acumen, and wisdom."

Enhancing brain function involves the adoption of a specific mental attitude that, according to Flynn, involves internalizing "the goal of seeking challenging cognitive environments—seeking intellectual challenges all the way from choosing the right leisure activities to wanting to marry someone who is intellectually stimulating. The best chance of enjoying enhanced cognitive skills is to fall in love with ideas, or intelligent conversation, or intelligent books, or some intellectual pursuit. If I do that, I create within my own mind a stimulating mental environment that accompanies me wherever I go."

James Flynn suggests the following example of how our social environment takes precedence over our IQ in determining how well our brain performs: Imagine a girl, let's call her Rachel, of average IQ born into a home where the parents make every effort to provide her with intellectually stimulating experiences. They buy her books, help her with her homework, hire tutors for subjects she finds especially difficult. Most of all, they do everything they can to ensure that she develops and maintains a positive attitude toward school. When problems arise with teachers or other students, the parents readily help to find a solution while being careful not to become overinvolved.

By dint of hard work Rachel is accepted into a superior high school, where she forms friendships with students who provide her with the intellectual stimulation she formerly received from her parents. Although Rachel has to work harder than the other students, she graduates in the top 20 percent of her class. She is accepted into a small but well-regarded university where, thanks to a class in political science taught by a gifted teacher, she decides to apply to law school.

Although Rachel's college grades aren't spectacular, because of the enthusiasm and dedication she shows during her interview, she's accepted into a law school. After graduation she joins a small firm specializing in intellectual property, her area of interest. Two years later, she marries a colleague who combines his legal career with a range of intellectual pursuits, ranging from playing in a jazz ensemble to taking postgraduate courses in Russian literature.

Note that in this example suggested by Flynn, Rachel's experiences have taken her further than her modest IQ would have predicted. Her environment outpaced her genetic inheritance as a determinant of her achievements. At this point her friends from school, her colleagues at work, her husband, and their friends provide the intellectual stimulation needed to keep her brain functioning at its best. None of this would likely have occurred if her parents hadn't encouraged Rachel to exercise her brain from an early age.

James Flynn's research underscores several important points. While it's true that genes play a role in determining intellect, social-cultural factors combined with effort play an even greater role. In other words, our cognitive functioning is largely modifiable. Each of the six traits associated by Flynn with enhanced mental abilities—mental acuity (fluid intelligence), knowledge and information (crystallized intelligence), memory, speed of information processing, curiosity, and the ability to think in abstract terms unrelated to specific applications—can be improved by one's own efforts.

Of these six cognitive processes, memory is the one that

most rewards our efforts at enhancement. A good memory is the bedrock of a superior functioning brain. A good memory also is a prerequisite for success in any aspect of life; learned new information can't be applied unless that information can be assimilated and related to earlier information and remain available for later recall. Anything that increases our memory powers increases our access to everything we learn. For these reasons, let's explore the ways memory can be improved.

Sharpen Sense Memory

Failures of memory—forgetting—result from two principle causes. The first is distraction. We can't recall someone's name moments after being introduced to him because during the introduction we were mentally preoccupied with something else and weren't really listening. Since we weren't really listening, our brain never successfully encoded the information and therefore cannot later recall it. The cure for this is straightforward: focus our attention on the name and the face during the introduction.

The second most common cause of forgetting relates to failures to register what is going on during the original experience. A fundamental way to enhance your memory, therefore, is to pay more attention to your sensory experiences. The more vivid your impressions about what you see, hear, smell, taste, or touch, the easier it is to establish a vivid and easily recollected memory. *So the first step toward an enhanced memory involves exercises in sharpening our senses.*

The director Lee Strasberg, founder of the Actors Studio,

taught his students to sharpen their acting skills by what he called sense-memory exercises. Harry Governick, one of Strasberg's early students, describes the first exercise, which involves an empty coffee cup:

After filling a cup with coffee, the student is encouraged to carry out every day for at least fifteen minutes a detailed exploration of every sensory aspect of the cup. Governick suggests starting with the sense of sight. "As your eyes view the cup, your mind should answer every detail about the visual aspects of the cup." Among other suggestions by Governick, the student should note the height of the cup, its diameter, its color, its material composition, and the dimensions of the cup's handle. He should look for ridges on the cup's lips, and note the shape and color of any artwork or ceramic design on the cup. He should also check for the shape and color of any reflections from the lights in the room that may be visible on the cup. "After you have exhausted every possible question your mind has asked your sense of sight to answer, move on to another one of the five senses, such as touch, and explore in the same deliberate exhaustive manner."

If the exercise is carried out faithfully, Governick claims, "you will actually 'see,' 'touch,' 'smell,' and 'hear' the cup and the coffee, as though it were right there in front of you. Your senses will faithfully re-create the cup and drink for you."

Although the sense-memory exercises were designed for actors to train their senses to respond on stage as they do in real life, I've found that nonactors can use the same exercises to sharpen sensory impressions and enhance concentration.

Visual sense memory exercises involve concentrating on

small familiar objects such as the cup suggested by Governick. Pick anything that appeals to you and study it intently. If your attention begins to lag, switch to drawing the object while saying out loud the features you are trying to incorporate into the drawing. Then describe aspects of the object you've noticed but cannot capture on paper because of limitations in your drawing skills. Similar exercises can be used to enhance acuity in the other senses.

As a *sound sense-memory exercise*, focus your attention on the ambient sounds around you. Do this in a relatively quiet area where different sounds can be distinguished easily. How many separate sounds can you hear? Identify them. Then concentrate on one sound at a time and write down all of the things that occur to you about it that distinguishes it from all other sounds. Or purchase a CD of bird calls and correlate the calls with the names and pictures of the birds. Other animal sounds will do just as well. For instance, when I discovered a large number of frogs living in and around the pond near my summer home in Prince Edward Island, Canada, I purchased a CD and guidebook to the calls of all of the frogs living in North America. At odd moments I play the CD and test how many distinct frog calls I can recognize. Admittedly, this isn't the most exciting of exercises; many people would find it downright dull. But the point is that it provides an intellectual sound challenge for me. So choose a sound sense-memory exercise that appeals to you.

More important than the sounds of birds, frogs, or other animals is the sound of human speech. Since human speech plays such an important role in our lives, it's useful to learn to detect subtleties in the sound of the human voice. An excellent test for auditory memory is to record a conversation and later attempt to mentally re-create everything that was said in the correct sequence. Novelists and psychotherapists are especially good at this. The novelist listens and remembers conversations because she needs to make certain the dialogue she writes captures how people actually speak—often not in complete sentences.

After I've finished recalling what was said and in what sequence, I check my recall against the recording. What did I forget? Does my recollection of the conversation occur in the precise sequence in which it occurred?

Next, I replay the recording, but this time I ignore meaning and content while listening to the voices as *simply voices*. I pay special attention to emotional features as signaled by changes in tone, cadence, loudness, the frequency of interruptions, and other changes in pacing. This exercise calls into play the right hemisphere of the brain, which is specialized to discern the nonverbal auditory signals whereby one person communicates with another.

For instance, several years ago I found myself uncomfortable and irritated when talking with a certain business acquaintance. Try as I might, I couldn't identify what it was about our conversations that bothered me. So I recorded one of our conversations and applied the two approaches mentioned above. In the first part—simply listening to the playback of one of our conversations and then writing it out from memory—I couldn't identify anything that explained my discomfort. Nothing in the dialogue seemed out of the ordinary.

But when I performed the second part of the exercise and concentrated just on the sound of the voices, I discovered an interesting and maddening conversational quirk.

Although my acquaintance never interrupted me, he nonetheless maintained control of the conversational space by means of a low peremptory hum that began whenever I started speaking. It was barely detectable—a kind of very subtle vocal "nudge" that continued throughout my part of the conversation. The overall effect was: "Please hurry up and finish what you're saying, because I want to speak once again."

Now I knew the source of my discomfort and annoyance: he never completely relinquished control of the conversation, even when I was speaking and he was supposed to be silent. But he was never silent, and if I hadn't recorded our conversation and listened to it as simply sound, I would never have discovered the source of my unease. Be prepared: at first you'll find it difficult to listen to your native language and hear it as just sound. That's because after learning language our brains are wired primarily for detecting the meaning rather than the sound. It's far easier to listen to a foreign language as pure sound because you can't process the meaning of the words you hear. But with practice you can pick up on tones of voice conveying emotion in your native language without consciously registering the meaning of what's actually been said.

Touch is the next sensation to be expanded. Start with fabrics like silk, leather, wool, suede, cotton, and cashmere. Arrange articles of clothing made from these materials on a sofa or a bed and, with your eyes closed, identify them by touch alone. Put them in your closet and over the next few days select a particular item by touch alone. As another touch sensory exercise, randomly set out twenty or thirty similar-size objects (a tennis ball, an orange, a potato, a parsnip, an onion) and sort them with your eyes closed, identifying each object by touch alone. Simply touch the object using the first three fingers of either hand. (Be careful not to pick them up or manipulate them in any way, since this provides information via receptors located in joints, tendons, and muscles.) Even better, ask someone else to select and arrange the objects; that way you won't know anything about the objects beforehand.

The next touch exercise is one I learned from a patient who worked as a blackjack dealer in Las Vegas. As she demonstrated for me on several occasions, she was able to cut a deck of cards and then correctly identify by touch alone the number of cards in each half of the cut deck. This skill required several years of regular practice to acquire, so don't expect to be able to accomplish it using a full deck of cards. Instead, try it with only a part of the deck (twenty cards is about right) Cut the cards and by touch alone estimate the number of cards in each half of the deck.

Some people's acquired skill at cutting cards by touch alone provides one of the reasons you should never play cards for money with strangers: a "shaved deck" may be in play. If a very thin layer of the upper outer edge of one side of a deck is finely shaved *with the exception of one card* (usually an ace), it's then possible to locate that card at any time simply by running a finger along that corner of the deck. The edge of the unshaved card juts ever so slightly from the shaved cards and can be detected by touch alone. To win, the person handling the deck has only to wait for a strategic moment and then draw that card from the deck or cut the deck so as to place that card on top.

Card sharks and pickpockets are masters of information management via touch alone. The pickpocket distracts his victim by touching one part of his body while deftly inserting a hand into the victim's pocket and extracting his wallet. And—shifting our attention to perfectly legal activities— magicians manipulate cards, coins, and other props by means of enhanced touch sensitivity in their fingers developed over hours of practice.

Touch can provide a wealth of interpersonal information as well. When you are introduced to someone, pay special attention to the steadiness and firmness of his handshake, the level of moisture on his palm, the smoothness or roughness of his hand. I do this every day as part of my evaluation of patients. If you have the opportunity to touch an article of clothing, identify it by its texture. Is that jacket made of real or faux leather? Is that dress silk or a synthetic blend? When you combine that kind of information with active sensory attention to the person's appearance and tone of voice, you already know a great deal about that person.

Tasting and smelling sensory exercises are as readily available as the nearest garden, spice rack, and wine-tasting group. Here is my favorite exercise: Take a number of spices (in their containers) at random and set them out on a table. Without looking at the labels or contents, by smell alone identify each spice. A recent selection included oregano, thyme, mint, minced onion, sage, sweet basil, cumin, chervil, orange peel, and black pepper. You may find several of them difficult and need to taste as well as smell the spice.

The linkage of olfaction with the brain is straightforward: thanks to the arrangement of the olfactory nerves and their connection with the forepart of the brain, scents such as perfumes or wines are experienced as complex and highly evocative perceptions. The process has been compared to the contributions of individual musical instruments to a symphony. Perfume-makers and winemakers routinely use terms drawn from music when describing their creations: top, middle, and bass "notes," each contributing to the harmonious "chord" of the scent. Just as a symphony cannot be any better than the performance of the individual artists and the quality of their instruments, smell (as well as taste) depends on each person's ability to detect certain basic odors. And this ability can be improved by practice. The best way to test your current olfactory ability is to have someone assemble seven small jars containing the substances mentioned in the next paragraph, then challenge yourself to identify them.

Eighty-five percent of the U.S. population can identify the following seven odors: baby powder, chocolate, cinnamon, coffee, mothballs, peanut butter, and soap. If you had trouble with one or two of them, don't worry. Surprisingly a few "normals" in the original test had difficulty identifying one or two of the odors.

Difficulty in smell identification is sometimes seen as an initial symptom of Alzheimer's disease. This finding suggests the intriguing though unproven possibility that Alzheimer's might be slowed in its progression by enhancing one's olfactory powers via scent identification exercises like I'm suggesting. (I'm not making that claim but at the very least such exercises challenge two of the brain functions affected by Alzheimer's: vocabulary, naming the scent, and memory, recalling past experiences with the scent.)

Next, I suggest *sensory (and motor) exercises* involving the hands. Since no part of the body is more functionally linked with the brain than the hands—with larger areas of the brain devoted to the fingers than to the legs, back, chest, or abdomen—developing nimble finger skills is a surefire way of improving brain function. Whenever you perform an activity requiring finger dexterity, you enhance your brain. Unfortunately, most adults—with the exception of musicians and surgeons—aren't highly skilled in fine finger control. To counteract this, take up a hobby that requires fine detail work such as knitting, model-ship or model-train building, bike repair, simple carpentry, painting, or drawing. A Pulitzer Prize–winning writer I know started taking drawing classes three years ago and now carries a sketch pad with him at all times. He's found that producing quick sketches of the things and people around him has sharpened his writing skills. Sketching helps him to notice additional details that he can then incorporate into his books and articles.

The final *sensory exercise involves body position and body shape:* changing what neurologists refer to as the *body schema.* In simple language, this term refers to the implicit knowledge that we all retain about the spatial properties of our body. For example, we're able to pass people on a narrow sidewalk without bumping into them because we quickly and unconsciously estimate the available space separating our arms and shoulders from them. We can do this because over the years our brain has built up a dynamic internal representation of our body and its extension in external space—in essence, the body schema. It's the integrity of our body schema that enables us to successfully return a tennis serve or sink that final putt during a hotly contested golf match.

Thanks to its dynamism, our body schema can be changed at any time by our own efforts. Usually these changes take a while before they are incorporated into the synthesis of a new body schema. If we opt for a new hairstyle or shave our beard, we're likely to be startled the first few times we pass a mirror and catch a passing glance at our new appearance. But after a few additional mirror encounters, our brain incorporates the stylistic changes into a new body schema.

The most fascinating aspect of the body schema is that it can include more than just our body. For instance, when we drive our car along a narrow street and pass another car with just inches to spare, we judge the available distance between cars because at that moment our car has become a part of our body schema. For some people this incorporation of their car into internal representation influences both their perception and their behavior. Researchers have found that the lines separating parking places appear narrower to the driver of a Humvee than they do to the driver of a Prius. And body schema considerations help explain why some drivers fly into a rage over a minor ping or dent on a car door—their body schema has been altered.

The fine-tuning of motor skills leads to the creation of new and more powerful brain maps. For instance, after years of practice, musicians incorporate their musical instrument into their brain's very circuitry. Eventually it's as if the musician and his instrument have become inseparable. Look at any of the many pictures of Itzhak Perlman sitting with closed eyes and intense concentration as he fondly cradles his violin. Or watch YouTube videos of the late Glenn Gould (playing Bach Partita Number 6 in E minor, "Toccata," or extracts from *The Art of Piano*), caressing the keyboard, his head lolling gently as he leans forward and whispers to his piano.

At times, this extension of body boundaries to include instruments and tools can take unusual, even amusing manifestations. Some people become personally attached to specific computers, cell phones, and other electronic gadgets. In some cases this leads to an insistence on the part of the owner that when the original unit malfunctions while under warranty, the manufacturer return to them the same unit rather than a replacement. That's one of the reasons many warranties now routinely contain such phrases as "The manufacturer reserves the right to repair or replace this unit at its discretion."

Before suggesting exercises for modifying the body schema, I want to describe an intriguing experiment involving what researchers refer to as peripersonal space (PPS), a force field that can be thought of as a virtual envelope around the skin's surface that extends our body boundaries.

In a variety of species, including humans, some brain cells respond to both touch and sounds occurring near the face or the hands. Think of an imaginary border around the head and the hands around which sound-touch integration occurs. As a result of this integration, we are quicker to respond to something touching the hand when it is accompanied by a sound. This multisensory PPS serves a protective function: in the presence of a loud sound, something touching the face or the hand is detected more quickly. For a blind person the PPS is even more critical than it is for a sighted person, because of the blind person's exclusive reliance for self-protection upon touch and sound.

According to research going back several years, the PPS of blind people is enlarged to include touch information coming not just from the hand but also from the tip of a walking cane. In essence their PPS has been expanded to form a new body representation where the space around the tip of the cane is as sensitive to touch as the space surrounding the hand. This holds true as well for those neurons that detect both touch and sound. Thus, if a noise originates near the tip of the cane, the blind person reacts to it as quickly as he would if it occurred close to the hand. In practical terms, such an arrangement enables the blind walker to detect touch or sound obstacles anywhere from the hand to the tip of the cane.

In an experiment carried out by a team of Italian brain scientists, normally sighted people received ten minutes of training in using a cane to find objects placed on the floor in a darkened room. That short training period was sufficient to alter their PPS to resemble that of blind people who regularly use a cane. They became as sensitive to touch and sound events occurring near the tip of the cane as to similar events happening near the hand.

In essence, just ten minutes of training with a cane by a person with perfectly normal vision brought about changes in PPS that were identical to those of a blind person. But this marvelous example of the plasticity of the adult brain was only temporary. Since the brain, thanks to its plasticity, changes from moment to moment, consistent and prolonged effort is required. That's why blind people must use their cane for many years in order to shape and maintain their augmented PPS. And since the sighted volunteers' training with a cane consisted of only a single ten-minute session, it should come as no surprise to learn that by the next day, in the absence of additional practice, their PPS boundary returned to the pretraining status: when blindfolded and holding their cane, the hand was once again much more responsive than the tip of the cane to touch or sudden noise.

The Body Has a Mind of Its Own, a 2007 book by science writer Sandra Blakeslee and Matthew Blakeslee (her son), contains a succinct description of the practical implications of enhancing one's peripersonal space:

> Your self does not end where your flesh ends, but suffuses and blends with the world, including other beings. Moreover this annexed personal space is not static. It is elastic. It morphs every time you put on or take off your clothes, wear skis or scuba gear, or wield any tool. When you eat with a knife or a fork, your peripersonal space grows to envelop them. Brain cells that normally represent space no farther out than your fingertips expand their fields of awareness along the length of each utensil, making them part of you.

Thanks to neurons in the frontal cortex that are bimodal (responding to touch and sound) or trimodal (responding to touch, sound, and vision), we activate identical brain areas whenever we use any of these three senses. Because of this intimate interplay among the neurons linking vision, touch, and proprioception, we're able through our own efforts to enhance our brain's functioning by integrating the information entering the brain from these senses. Trained athletes and musicians establish this integration through long years of practice leading to the development of a keen kinesthetic sense.

For example, a baseball pitcher can begin his windup and at the last second change direction and fire the ball to his left in a bid to pick off the first-base runner. Or a basketball player while charging down the court with the ball can look to his right in a feigning maneuver, change direction in an instant, run toward the basket and score two points with a hook shot. In both examples the athlete coordinates vision (what's seen) with proprioception (what's felt).

Even though most of us aren't aiming at achieving the sensory integration of the athlete, we can enhance our brain's functioning to enlarge our PPS by integrating our vision, sound, touch, and joint senses.

As a preliminary exercise in extending peripersonal space, stand with your eyes closed and your hands at your sides. Now raise your arms to the horizontal position where they are exactly level with your shoulders along a straight line across your back so that the tip of the middle finger of one hand is on alignment with the tip of the middle finger of the other hand. Now extend the forefinger of each hand and close the other four fingers into a fist. Then, in a slow, sweeping, embracing motion in front of you, move the hands toward each other until you judge that the forefingers are just about to meet at the midline directly in front of you. At that point, stop moving your hands and open your eyes. Most people find that their forefingers are not on a direct projectory toward each other, but are off course by as much as an inch. In some cases, one or even both of the fingers will no longer be pointing in a horizontal direction, but slightly up or down. Repeat until you get it right.

When you've mastered that preliminary exercise, sit barefooted in a chair with your right leg fully extended in front of you. Close your eyes and pretend that your right hand is a gun with the forefinger forming the barrel. Now with your eyes closed, point the forefinger directly at your big toe. When you're confident that you've positioned the barrel of your imaginary gun so that you will be able to "shoot" your big toe, open your eyes. Few people are able on the first try to form a straight line from forefinger to toe in this test frequently used by neurologists to test joint (proprioceptive) sense. Joint-sense estimation is especially inaccurate in people with diseases affecting the nervous system anywhere along the route from the peripheral nerves in the legs to the parietal lobes, where joint and position sense is ultimately processed. But even people with normal nervous systems initially find it hard to point directly at their toe with their eyes closed. That's because we rely primarily on our eyes to relay information to the brain. When we depend on sound, touch, or position sense we don't do as well. But that overdependence on vision can be balanced by exercises that enhance PPS.

Tai chi provides one of the most effective long-term

approaches to enhancing peripersonal space. This ancient dancelike exercise requires the practitioner to perform a series of slow motions while simultaneously focusing attention on specific body areas, especially the hands and tips of the fingers. Over time this leads to alterations in the body image, especially PPS. For example, experienced tai chi practitioners develop a tactile acuity in the fingertips rivaling that of musicians and blind Braille readers. This suggests to C. E. Kerr and his Harvard-based colleagues that tai chi may create a plasticity within the brain similar to that found among individuals who play musical instruments, read Braille, or engage in other activities that require finely honed fingertip sensitivity. "There is a strong connection between tactile spatial acuity at the fingertips and measures of brain function," says Kerr.

In order to get the maximum brain benefit, I suggest that you find a well-trained tai chi teacher and take some lessons. If this isn't practical, several excellent DVD instructional programs are available. But whichever path you choose, you'll learn as you move through the tai chi form, progressing from one of the sixty or more positions, that the hands and feet must be correctly and precisely aligned. To help you do this, your instructor will periodically ask you to pause, close your eyes, and mentally form an image of the current positions of your hands and legs. After you have envisioned the precise position of your hands and feet, he'll ask you to open your eyes and check how accurately your internal image corresponds to your body position. It takes years of practicing tai chi forms before the internal image and the external positions correspond exactly.

Other everyday examples of extending peripersonal space include alternating between a pen and a word processor; learning to play a musical instrument; purchasing a newer and lighter tennis racket that frees your swing; spending a weekend at an auto-racing camp where you can learn high-speed control skills as taught by competition race-car drivers. In these instances the pen, the musical instrument, the tennis racket, and the high-performance car extend and enhance the body's performance by creating new functional maps within the brain. The experience is similar to what happens when the blind person becomes skilled in the use of his cane. The more time spent creating these new maps the greater your proficiency. And while you can follow my suggestions of how you can form these new maps the best approach is to come up with your own program that incorporates activities that are interesting to you. The key is to activate those neurons that construct our PPS.

After enhancing sensory memory by performing one or more of the suggested approaches, you're now in a better position to work on *ways to improve memory in general.*

Augment General Memory

Science fiction writers are fond of presenting their readers with characters who no longer know who they are because their memories have been obliterated. In the 2000 movie *Memento*, the main character, Leonard Shelby, lives in an anxious and forebodingly eternal present; he can't recall any of his past experiences and as a result doesn't know who he is.

But such devastating memory failures aren't just the stuff of fiction.

A real-life patient (identified as H.M. in order to protect his privacy) lost his capacity to form new memories following a brain operation that removed the hippocampus and parts of his anterior temporal lobe on each side of his brain. As a result of his absent hippocampi (the plural form; there is one hippocampus on each side of the brain), H.M. permanently lost the ability to form new memories. Each time he encountered his doctor the experience was a new one for him. His memory was so bad that after learning of his mother's death, he soon forgot this distressing information, and over the years he became upset and wept anew whenever anyone mentioned that she was no longer alive.

Despite its tiny size, the hippocampus is one of the most powerful structures of the human brain. Everything you ever learn wends its way through this portal prior to its distribution throughout the remainder of the brain. If your hippocampus is damaged as a result of a stroke or traumatic brain injury, you can no longer establish new memories, as happened with H.M. A similar but less severe failure occurs with depression. As part of the depressive experience, the hippocampus atrophies and memory failures become commonplace, often leading to an incorrect diagnosis of dementia. But whatever the cause, impairment of the hippocampus causes newly learned information to drift off a person's mental horizon after only a few moments. What's worse, this drift into mental oblivion can't be prevented despite a person's best efforts.

In my office I routinely test hippocampal function in those

patients I suspect may be suffering from damage in that area. I ask them to learn and retain five simple words, such as *apple, tie, pen, house,* and *car.* Even severely affected patients can initially repeat the words back after a few attempts. The problem arises when I ask them to repeat the words again after a five-minute delay; those with hippocampal failure can't remember more than one or, at best, two of the words. Persons with normally functioning hippocampi, in contrast, can usually manage to come up with all five of the words. People with damage to the hippocampus perform so poorly because their brain can't encode the test words—a prerequisite for holding them in short-term storage.

According to recent research, both the recollection of past events and the imagining of future ones require a normally functioning hippocampus. For instance, one fMRI study of healthy volunteers revealed that recalling past experiences and imagining future experiences activates a similar network of brain regions, including the hippocampus. As further proof of a memory-imagination link, people with faulty memories secondary to dysfunctional hippocampi have difficulty imagining anything that hasn't actually happened. For example, those with hippocampus-related memory problems do poorly when imagining a meeting with a friend on a specific street corner or dining with a friend in a familiar restaurant. It's as if both the imagined and the recollected world are rendered in shades of gray instead of vivid color.

This new discovery of a link between memory and imagination was anticipated in Greek mythology where Mnemosyne, the goddess of memory, gives birth to the muses, a sisterhood of spirits who inspire the creative process through remembrances. It also suggests a *strategy: the use of memory exercises as a means to foster imagination and vice versa.* Creative-writing teachers have always had an intuitive feeling for this approach. They encourage their students to provide from memory highly specific details about the places and people they have known. This enables the writer to create in his or her imagination more memorable scenes and characters. Using this technique, the author can render his fictional creations with greater believability since they're based on real people and places that he remembers from previous encounters. In order to provide remembered details about real places and people, the writer calls upon the hippocampus.

Since strengthening memory also increases imagination and creativity, it's worth the effort required to develop a superior memory.

Let's start with the simplest memory exercise of all: repeating something that has been heard just moments earlier.

Forward digit span: Ask a friend to read to you in succession strings of numbers, starting with five-digit spans and working upward (six digits, seven digits, etc.) until you can no longer correctly repeat back the number sequence. He or she should say the numbers in a steady monotone, taking care not to group the numbers. When they've finished, repeat the numbers back in the precise order.

If you prefer to do the exercise by yourself, read in succession five-, six-, seven-, and eight-number strings into a recorder and a few hours later listen to that recording of yourself reading the number strings one at a time. After hearing each number string, say it aloud and write it down. When you're finished, reverse the recording and listen to your response and check it for accuracy. If you correctly recited the five-number string in its precise sequence, go on to a six-number sequence, then seven, then eight, etc.

As an alternative, you can test visual rather than auditory recall by looking at number strings instead of listening to them. Write out the numbers zero through nine on separate two-by-two-inch pieces of paper and paste each of them onto cardboard or plastic. Place the numbers facedown on a table, select five of them, and then turn them over and randomly line them up in a straight line. Look quickly at the numbers and then turn away and write them down from memory. Look back at them and check your accuracy. Then proceed to six-, seven-, eight-, and nine-number strings.

When listening to the numbers and repeating them back, the longer your digit span the better your auditory processing. The same holds for visual processing: the longer the string of numbers you can write down after reading them the better your visual processing. As a side benefit, comparing your auditory and visual performances provides a quick and informal way of determining if you are primarily a visual or an auditory learner. In addition, forward digit span—either listening to or reading the numbers—is an excellent means of enhancing focus and concentration. Even more important, it's possible to increase one's digit span by practice.

It may surprise you to learn—as it did me—that the average digit span for normal adults is only between five and seven. A two-year-old has a digit span of two, a three-year-old a span of three, a four-year-old a span of four, etc., up to a span of seven for a seven-year-old. In other words, the average digit span for adults is set by age seven—and this holds true not just for English speakers but also for native speakers of other languages. It seems that the brain isn't designed for spontaneous encoding into short-term memory of more than seven items. If you want to do better than that, you have to practice.

At this point you might wonder, "Why would anyone bother? So what if I can learn to increase my digit span?"

Despite its simplicity, digit span reflects the efficiency of the earliest stages of information processing within the brain. This is important because how well you learn depends on how efficiently you process information at the earliest stages (the basis for short-term memory). Heightened attention enables you to increase the amount of information that your brain can take in (as mentioned in the discussion of sensory memory). If you increase your digit span, you can improve your brain's performance. We know this because digit span has been found to be a reliable predictor for early math and reading proficiency. The same holds true for performance on standardized tests.

Nor should it be surprising that increases in digit span correlate with enhanced brain performance. Among the brain functions activated are attention, concentration, sequencing, number facility, and auditory and visual short-term memory.

Backward digit span: Repeat the previous exercises, but this time recite the numbers in reverse order. If you're doing the exercise by yourself, it's especially important to record your responses and check them for accuracy, since it's easy to make errors toward the end of the backward digit span exercise (when you are recalling the first few numbers of the sequence) and not be aware of your failure.

If you carry out both the forward and the backward digit span exercises, you'll appreciate how much harder it is to mentally "read" the numbers in reverse. While many people can accurately recite seven digits forward, few people can manage more than five digits backward. Try it yourself. The reason for that difference? Backward digit span involves not only the registration and encoding of numbers but also their manipulation. You have to envision the numbers in some way—depending on the memory technique employed—and then "read" them from right to left rather than following the usual practice of reading the numbers from left to right. In addition, different brain functions are involved. While attention and auditory short-term memory are called into play with forward digit span, attention and working memory are used in backward digit span.

Working memory is the ability to hold information "online" for later retrieval while turning your attention to something else. For example, when you're on the phone and momentarily pause in your conversation in order to answer a "call waiting," it's your working memory that enables you to pick up where you left off when you return to the original call. Similarly, when you reverse the digit span, moving leftward from the last number to the first, you must hold in working memory the first few digits in the string in order to repeat them at the end.

Similar difficulties occur when reversing words. Mentally spell the word *world* backward. You should have no problem doing that. Then spell *hospital* backward. If you did that correctly (be certain to use your voice recorder to check for accuracy since the *p* and the *s* are frequently transposed), try *administrator,* then *equestrian.* Most people can manage *hospital* and *administrator* but flounder when backward spelling *equestrian.*

Notice that *administrator* contains thirteen letters, eight more than the five-digit upper limit for numbers that most people can comfortably reverse. There are two main reasons why we're better at reversing letters than numbers. With the exception of mathematicians, most of us spend more time working with letters and words than we do with numbers. In addition, words often represent something while a string of numbers is meaningless—unless we can impose a meaning by using one of the memory techniques described next.

Cultivate the Art of Remembering

Obviously, few of us spend much time during our everyday lives memorizing lists of digits and repeating them, either forward or backward. Most of the time, we're learning new information from other people, e-mail, the Internet, or other sources. This is the kind of memory we're enthusiastic about improving. Further, cultivating the art of memory can balance our overreliance on written sources of information.

In Plato's *Phaedrus,* Socrates recounts the judgment of the Egyptian god Thamus on the invention of writing by the god Thoth: "This invention will produce forgetfulness in the minds of those who learn to use it, because they will not practice their memory. Their trust in writing, produced by external characters which are not part of themselves, will discourage the use of their own memory within them."

Although it's a long trip from the early Greeks to the BlackBerry, Socrates' warning remains relevant today. As we become more dependent on memory aids, our natural mnemonic powers are atrophying from disuse. And this leads to a vicious circle: as memory deteriorates, we rely on storing more facts on paper or, as is increasingly common, on electronic devices, thus further weakening our memory powers. If you want to think smarter, you have to make active efforts to strengthen your memory and reverse this process.

All techniques to enhance memory involve the imposition of *meaning* on the information we wish to remember. For instance, memorize the following list of numbers by reading it aloud just once, then look away and try to recall the numbers in their exact order: one, four, nine, one, six, two, five, three, six, four, nine, six, four, eight, one. Most people can manage about seven of the fifteen numbers. You can, however, recall all of them if you first convert them into single digits (1, 4, 9, 1, 6, 2, 5, 3, 6, 4, 9, 6, 4, 8, 1) and observe that they can be grouped to represent the squares of numbers one through nine: 1, 4, 9, 16, 25, 36, 49, 64, 81.

Another way of increasing your recall is to focus on the meaning of something you are trying to memorize. In a famous experiment carried out in the mid-1970s, volunteers were asked one of three types of questions about a series of words they were requested to memorize. One question concerned the word's visual appearance ("Uppercase or lowercase letters?"). A second question focused on the sound of the word ("What does it rhyme with?"). The third type of question probed the meaning of the word. When later asked to recognize the words from a list, those who had encoded the words according to meaning greatly outperformed those who encoded according to visual appearance or sound.

Various methods have been suggested over the centuries to improve memory by meaningfully encoding the material to be memorized. Most of them can be traced to an unknown teacher in Rome who around 86–82 B.C. compiled a list for his students of specific techniques.

According to this teacher, who described his method in the *Rhetorica ad Herennium*, the "art of memory" involves skillfully linking the items to be remembered with places and images. If you want to remember a grocery list using this method, you create a mental picture of a familiar room and then place the objects you want to remember at select locations within the room. You then imagine yourself walking through the room and observing the objects at the locations where you placed them. You can rapidly confirm the power of this method for yourself.

Think about your living room. Envision as many details as you can about it. Then mentally place the following objects at various places in the room (above the fireplace, on a favorite chair, on the coffee table): log, candy, wallet, comb, pen, tie pin, ring, watch, and shoe. Take care to clearly envision the object at its "locus" (site) within the room. The clearer your image, the more accurate your recall will be.

Alternative methods employ the story system. Imaginatively weave together into a coherent story the items to be remembered. The more vivid, colorful, novel, and even bizarre the story, the greater your chances of remembering your list. For instance, for the list mentioned above, imagine a log dressed like a man walking along the street eating candy. The log is quite dapper and is decked out in an expensive suit and Bruno Magli loafers. His solid 18-karat-gold accessories (tie pin, ring, pen, and watch) gleam in the sunlight as he smugly struts along, the quintessence of self-satisfaction. But all is not well. Sneaking up behind him is a pickpocket pretending to be combing his hair while slipping his hand in the log's back pocket, from which he extracts a fat wallet. If you fix that image clearly in your mind, you'll be able to recite the list an hour, a day, or a week from now. That's because the story links the words into a vivid quirky image that is easy to remember.

The same principle holds true if you want to remember numbers. For most people, numbers are harder to remember than words simply because mentally "seeing" numbers isn't as easy as envisioning objects. While it's easy to imagine a shoe, a watch, and a pen, how do you mentally envision the number 313 or 4,573? There are several ways of doing that.

When I asked puzzle master Scott Kim to quickly reverse the seven-digit span 0785621, he responded within a second or so after I finished reading the number to him. Since he knew that the string would contain seven numbers, he broke it down into a four-digit number, 0785, separated by a space from the three digits of 621. He then simply reversed the three-digit number to yield 126 and the four-digit number to yield 5870.

"If the digit span was longer, I would have probably associated the higher numbers with a higher musical pitch and the lower numbers with a low pitch," Scott explained to me. "Or I might try something else and imagine the numbers occupying places on a grid. Or I might use my kinesthetic sense and imagine myself touching the numbers, with each of the numbers having its own special 'feeling.' "

Others have favored additional strategies. Memory wizard Rajan Mahadevan devised a system based on his memorization of telephone area codes. When memorizing 212 he associated it with New York City, the sequence 202 he associated with Washington, D.C., 410 with Baltimore, and so on.

My personal favorite method for remembering numbers is to convert them into images and then recall the numbers by reconverting the images to numbers. Here's how to do that:

Memorize the following images that employ a rhyming word for the numbers from 0 to 10.

0. Hero
1. Sun
2. Shoe
3. Tree
4. Door
5. Hive
6. Sticks
7. Heaven
8. Skate
9. Vine
10. Hen

Because of the rhyming you'll find the list easy to memorize. Each time you see a 0, imagine Leonardo DiCaprio or your favorite movie hero. When you see 7, you can imagine an angel floating on a cloud and playing the harp. If you don't fancy that image, picture a loaf of bread; *leaven* is an alternative rhyming word for *seven* and brings to mind a loaf of (leavened) bread. The other eight numbers are easier to remember because they involve universal images.

Take the number sequence 2023627834 and memorize it using this system. Here's one way of doing it. Group the first three numbers into the image of Leonardo DiCaprio holding a shoe in each hand (202). In order to remember 36, picture a power saw reducing a tree to a pile of sticks. For 27, imagine an angel playing a harp in the form of a shoe. For 834, envision a skate coursing along an icy pond shaped in the form of a tree and crashing into a barn door. Other images will work just as well. In fact, you are more likely to remember images that you come up with yourself because in the process of composing them you will be utilizing *association* and *elaboration*, the two most basic memory techniques.

Whenever you attach a word or a name to something familiar, you are practicing association (associating a face with a name, for example). Sometimes the association comes easily. My dermatologist is named David Spot; the most famous neurologist in the mid–twentieth century was W. Russell Brain. No problem remembering either of them. Other names, such as that of Dr. Paula Trzepacz, my successor as president of the American Neuropsychiatric Association, aren't as easily reduced to an image.

Elaboration is another basic memory technique. The more meaning you can give to the thing to be remembered, the more successful you will be in recalling it. One way of doing this is to devise a sentence that can work as a code for recalling the number. A famous example is the sentence devised by the British mathematician Sir James Jeans to represent pi, the ratio of a circle's circumference to its diameter. "How I want a drink, alcoholic of course, after the heavy lectures involving quantum mechanics!" The number of letters in each word in that sentence when read in sequence corresponds to the first fourteen decimal places of pi: 3.14159265358979.

In medical school I employed a similar technique to learn the names of the twelve cranial nerves by memorizing this sentence: "On Old Olympus's towering tops a Finn and German vied at hops." The first letter of each word corresponds to the first letter of each cranial nerve (olfactory, optic, oculomotor, trochlear, trigeminal, abducens, facial, auditory, glossopharyngeal, vagus, accessory, and hypoglossal).

What do these various memory "tricks" have in common? Each method forces the memorizer to focus attention on the information to be memorized. This is the key requirement for the successful recall of a name, for instance. Most failures in name recall result from inattention rather than a memory failure. As mentioned earlier, the person who cannot recall a name at a later point after an introduction probably wasn't really listening to the name during the introduction. Since she didn't really hear it, she can't later remember it. Listening closely, asking for a repetition of the name, inwardly repeating it, and finally trying to come up with an image to link name and face—these steps will greatly increase the likelihood of later recall.

Since we are visual creatures on the basis of our brain's

organization (the number of fiber tracks conveying visual information within the brain far surpasses the number of fiber tracks conveying any other form of sensory information), we increase our chances of remembering something by converting it into a visual format. The odds of correctly recalling the name can be further improved by using stark and even bizarre images. Memory savants differ from the rest of us as a result of their astounding ability to convert information of any type into pictures.

A useful method for remembering textbook material is encoded in the mnemonic PQ4R, which is shorthand for the steps involved. First, *preview* the pages to be remembered by quickly scanning through the chapter and noting the headings. Next, compose *questions* about what you've read. Then perform four Rs: *read* the text in search of answers to those questions; *reflect* on what you've read and try to come up with examples of how you can apply it; *recite* the material after you have learned it. The easiest way of doing that is to put the book to the side and recall and recite what you have just read. The last step in PQ4R is to *review*: try to recall and summarize the main points. PQ4R will work even better if you use a voice recorder. That way the information enters the brain through both the visual and the verbal pathways.

I regularly use the PQ4R method and can vouch for its effectiveness. I prepared for my board examination in neurology by quickly reading background books, formulating questions about what I had read, and speaking those questions into a voice recorder. I then searched for answers in the text, reflecting out loud into the recorder the implications of what I had just learned for how I would care for my patients. Next,

I recited what I had learned and recorded it. Finally, I sum-marized and reviewed everything aloud. At odd moments over the next few weeks I either listened to the recordings (while driving) or during more leisurely moments I reviewed the written material while listening to my voice summarizing it on the recording.

In order to proceed further in strengthening your memory, I suggest two formal memory courses that you can do at your own pace. A short beginner's course can be found at www .NeuroMod.org, the website of the Neural and Cognitive Modeling Group at the University of Amsterdam. It provides a discussion of the basic principles by Professor Jaap Murre. When visiting that site, click on "Human Memory" and select "online memory improvement course."

You'll find a more rigorous course at Kevin Jay North's www.Thememorypage.net. When visiting that site, click "Tutorials" and select "How to Improve Your Memory." This includes novel techniques for memorizing lists and numbers as well as instructions on the use of peg words. I found most useful his tips on increasing one's reading retention—you can find it by clicking on "Tips&Tricks" from the main menu. For really advanced memory techniques, North provides a free link to the Memory Master Training Course, essentially a book-length presentation of memory techniques. There are also some strictly-for-fun applications such as a card-counting technique coupled with an online test of your skill after you've learned the method.

Is There an Outer Limit to Memory?

After improving your memory by using one of these methods, you are probably curious enough to ask, "What is the greatest memory performance that I can reasonably expect to achieve?"

In search of an answer to that question, I watched Morley Safer's 60 *Minutes* interview with Daniel Tammet, one of an estimated only fifty true memory savants living in the world today. Thanks to his wire-framed glasses, prominent front teeth, and close-cropped hair, Tammet fits most people's image of a savant. But Tammet differs from the typical savant in several ways. He isn't retarded but has Asperger's syndrome, a form of autism marked by awkwardness and difficulty in social situations coupled with sometimes bizarre talents and interests. For instance, people with Asperger's can be found within the ranks of compulsive hoarders and collectors, especially collections involving odd items that are of little interest to the average person (bottle caps, for example). Often a person with Asperger's displays superior intelligence in selective areas, such as memory and mathematics. Finally, their thought processes are marked by prominent defects in imagination and empathy.

During the interview Tammet appears shy but likable, engaging in a relaxed conversational repartee marked by a social facility and occupational success that sets him apart from the typical person with Asperger's. He runs a successful Web-based business for language tutorials.

As I watched Tammet describe the virtuoso memory

performance that has made him world famous, two things occurred to me. First, the memory performance of a savant like Tammet isn't simply an extreme example of what we all could do if we simply stopped lazing around and put in additional effort. If you need convincing about this, try reciting as many as you can of the numbers of pi. As a convenient round-off in trigonometry, most of us learned to use 3.14 as a working approximation of pi. But pi is an infinite number with the digits to the right of the decimal point extending into infinity. "It isn't possible for someone to write down the number pi exactly, even if he or she had a piece of paper as big as the universe to write it on," writes Tammet in his autobiography, *Born on a Blue Day*.

Even though pi is infinite, that hasn't stopped memory virtuosos from competing with each other for the distinction of reciting the longest string of its digits. Various techniques have been used. As mentioned earlier, anyone can easily recite pi to fourteen places by converting numbers into a sentence with the words composed of letters corresponding to the numbers. Others convert the numbers into familiar patterns that vary according to interests and life experiences (athletic performance statistics, historical dates, zip codes) and enable the mnemonist to store extended number sequences in long-term memory. By using a conversion process, the Indian mathematician Rajan Mahadevan in 1981 correctly recalled the first 31,811 digits of pi.

On March 14, 2004—International Pi Day—Tammet accepted the challenge of exceeding the European record for reciting from memory as many digits of pi as possible (22,500 digits was the European record at that time). Tammet's method, however, involved more than a simple conversion process:

"When I look at a sequence of numbers, my head begins to fill with colors, shapes, and textures that knit together spontaneously to form a visual landscape. To recall each digit, I simply retrace the different shapes and textures in my head and read the numbers out of them."

As this comment makes clear, Tammet based his performance on synesthesia, specifically a form known as grapheme-color synesthesia, where numbers and individual letters of the alphabet (collectively referred to as graphemes) are perceived as colored. "I see numbers as shapes, colors, textures, and motions," Tammet said. He also uses number-form synesthesia: a mental map that automatically and involuntarily springs to mind whenever he thinks of numbers. Thus the numbers don't line up along a horizontal line but, as Tammet describes it, "as the sequence of digits grows, my numerical landscapes become more complex and layered, until—as with pi—they become like an entire country in my mind, composed of numbers."

Over five hours and nine minutes, Tammet recited 22,514 digits to beat the earlier British and European record. "As I recited I could feel myself becoming absorbed within the visual flow of colors and shapes, textures and motion, until I was surrounded by my numerical landscape."

I think it's fair to say that Tammet's memory technique isn't an option for those of us who aren't synesthetes. But we can still improve our memory by following his advice to cultivate "single-mindedness along with a strong drive to analyze detail and identify rules and patterns." Here are several rules that will help you do that:

1. *Pay attention.* Mentally focus on the information you want to learn to the exclusion of everything else.

2. When encountering a new piece of information that you want to remember, *employ as many sensory faculties as possible.* Listen to yourself speaking the name or fact; write it down; read aloud what you've written; finally, with the index finger of your dominant hand, write it on the palm of your other hand. This sequence funnels the information through the visual, auditory, and tactile channels.

3. *Put the information you want to remember into an image,* preferably one that is captivating because of its novelty and high-image content.

4. *Create your own memory aids based on the details of your life experience.* The goal is to capture unfamiliar information by linking it with images of things that are unique to you, e.g., the layout of your living room or other familiar location. With this method you have only to place the information you want to learn at key locations within the room and then retrieve it by mentally strolling through the room.

5. *Learn a list of memory pegs* that can be used for numbers. One example is on page 105, but you can easily come up with your own associated rhyming words.

Finally, a few words about forgetting. Why do we forget? "Memory decay" over time was a formerly popular explanation. But blaming everything on the passage of time isn't a satisfactory explanation. It's like explaining rust on a car as resulting from the passage of long periods of time. Rust results not from the passage of time but from the oxidation processes that occur in the metal. Like rust, forgetting requires something in addition to the mere passage of time. Forgetting as a simple by-product of time-linked decay is also contradicted by the fact that sometimes memories recover over time, or can be resurrected by reminiscence, psychotherapy, or other techniques.

Currently the most popular theory of forgetting is *retroactive interference:* new information interfering with and supplanting older information. For instance, do you remember what you had for lunch five days ago? Unless the lunch was a special occasion, you're unlikely to be able to distinguish it in your memory from all of the other lunches you've had over the past week. The more active your life, the more difficult it will be to overcome retroactive interference. Which leads to a marvelously effective memory exercise: remembering similar situations by mentally linking them via distinguishing characteristics, however subtle. For instance, even though the food may have been unremarkable at any particular lunch, the discussion or the information gained or some other aspect of that lunch can make the total experience—including what you ate—stand out in your memory. The key is to discover or create that distinguishing feature. Let's take up a few ways of doing that.

Improve Working Memory

Have you ever wondered why telephone numbers are seven digits long? Because, as first pointed out by the nineteenth-century philosopher Sir William Hamilton, people generally experience difficulty with numbers containing more than seven digits: "If you throw a handful of marbles on the floor, you will find it difficult to view at once more than six, or seven at most, without confusion."

Hamilton's homespun experiment was confirmed in a 1956 paper by Princeton University psychology professor George Miller, "The Magic Number Seven, Plus or Minus Two." Miller found that those who exceed that seven-digit span do so by breaking the number into groups of smaller numbers of recognizable sequences or words, a process he called "chunking." For numbers exceeding seven digits, chunking is an indispensable tool. Even seven-digit numbers such as telephone numbers are broken up into two "chunks": the first three numbers interrupted by a pause (for spoken numbers) or a space (for written numbers), followed by the last four numbers—the memory method used by Scott Kim described earlier. For international calls, several additional chunks must be memorized.

Chunking makes it easier to keep all previous numbers "in mind" as each succeeding number is encoded within the brain. For instance, when someone tells you a telephone number, you won't be able to dial it unless you can keep in mind the correct sequence from start to finish. This process of encoding one item while retaining access to items encoded moments earlier depends on *working memory*. It is as a rule considered the basis for general intelligence and reasoning, since people who can hold the greatest numbers of items in mind are best equipped to consider multiple aspects of a problem simultaneously.

Word processing on your PC provides a good analogy for what happens during failures of working memory. When you switch from one document to another on your word processor, the unattended document is still accessible. All you have to do is toggle from document 1 to document 2 in order for you to keep both of them "in mind." But if you close one document as you move to the second document, that first document is no longer available for immediate retrieval. A failure in working memory is like that: you close the first document when you switch to another instead of holding that first document online.

For instance, if you're interrupted by an intercom message from your secretary at the moment you're about to make an important point with an employee during a meeting in your office, you may discover after taking that message that you've forgotten what you had intended to say before the interruption. This "interference effect" results from a failure in "working memory." You inadvertently closed the first document (the point you were intending to make during your in-office discussion) when you moved to the second document (the intercom message from your secretary). As a result, when you returned your attention to the office conversation and attempted to retrieve what you intended to say before the interruption, you drew a blank.

The greatest enemy to people's working memory is distraction. If you're thinking of something else and aren't really listening when you are told a telephone number, the sequence won't be encoded in your brain. Later, you won't be able to retrieve it because distraction during the initial encoding process interfered with memory consolidation.

Multitasking, despite its inefficiency, has the potential to strengthen working memory as long as people make a deliberate effort to keep the multiple tasks "in mind." As the skilled multitasker switches from one activity to another, if he's adept enough he retains the first activity in working memory. As a result, in contrast to the example of the interruption during the meeting, the skilled multitasker has no problem remembering the point he was about to make just prior to the interruption.

Working-memory deficiencies underlie the distractibility and poor academic achievement of people, adults as well as children, who have attention deficit/hyperactivity disorder (ADHD). Reading problems also result from working-memory problems: the earlier words in a sentence can't be recalled by the time the reader reaches the end of the sentence.

Many of the classic memory exercises are actually exercises of working memory. Here's an example of one that I use in my office to test my patients' working memory: Memorize these four items: *apple, charity, Mr. Johnson, tunnel*. Now set a chronograph or other timing device for five minutes and return to your reading. When the alarm goes off, recite aloud the four items. To do that you had to keep the items in a kind of suspended animation: available for recall and yet not so much "in mind" that they interfered with your ability to understand the pages that you read before the sounding of the alarm signaling that you should recall the items.

In order to distinguish working memory from memory in general, think of it this way: If you're required, as in the above example, to hold certain items "in mind" while you turn your attention to something else, you're using working memory. If you're simply asked to recall something from the past ("Who was the sixteenth president of the United States?"), that is a test of general memory. Working memory is also called upon whenever we mentally manipulate information. For instance, memorize these numbers: 238538392. Now, as an exercise in working memory, rearrange that sequence from lowest to highest (223335889) without writing anything down.

As another example of the general memory–working memory distinction, recall the names in order of all of the presidents since Harry Truman. That is an exercise in general memory. Now with the list firmly in mind, rearrange the names in alphabetical order without resorting to pencil and paper (Bush, Bush, Carter, Clinton, Eisenhower, Ford, Johnson, Kennedy, Nixon, Reagan, Truman). In this exercise you're exercising working memory to mentally carry out the rearrangements.

Admittedly, sharp distinctions can't always be made between general memory and working memory. Often we call on both kinds of memory at the same time. For example, if you're aware of all of the cards in play during an ongoing game of bridge, you're exercising your working memory; if you're thinking about how things went when you held a similar hand in a game you played several months ago, you're using general memory.

Working memory and *general memory* also involve different brain pathways. Information held in general memory is initially encoded in the hippocampus for later distribution to the temporal lobes and other portions of the two hemispheres. And since general memories are distributed throughout the brain—rather than encoded within a "memory center"—forgetting is a gradual rather than a sudden process.

For instance, you don't suddenly forget all of the details of your college graduation. Instead, with each passing year your general memory becomes less precise and reliable. That's because the memory of your graduation is widely distributed throughout your brain rather than localized to a specific area. This widespread distribution has led psychologists to compare general memory to a hologram, a laser-generated three-dimensional image.

Holograms differ from ordinary photographs by nature of the fact that the entire image can be reconstructed from portions of the hologram. If you divide a hologram into two or more pieces, the resulting image depicted on the parts become less distinct. It doesn't disappear altogether even when the original photo is divided many times. General memory is like that too, because of its wide distribution throughout the brain.

Working memory, in contrast, is formed at a later point in the information flow. After encoding in the hippocampus and wide distribution throughout the rest of the brain (principally the two hemispheres), working memory is maintained in an active state within the frontal lobes, specifically the dorsolateral prefrontal cortex. As working memory improves, the rate of glucose metabolism *decreases*. The explanation for this decrease?

The decrease in brain activation that is associated with improved performance suggests that with practice our brains don't have to work as hard to achieve improved performance. The process can be compared to learning to drive a car. The more experienced you become in driving, the less conscious attention and concentration you have to exert. On the basis of your previous driving experience, you're able to monitor and respond to nearby traffic without thinking consciously about how you're doing it. An fMRI of your brain taken while you're driving would show decreasing activity in the dorsolateral prefrontal cortex as your driving skills increase. A similar decrease in brain activation occurred in an experiment involving volunteers learning to play a computer game. Initially their prefrontal glucose metabolic rates were high. But after they became proficient at the game, their scans showed significantly lower rates of glucose metabolism.

Since working memory is carried out primarily by the frontal lobes, and these areas are especially vulnerable to atrophy because of disuse as we age, it's important to do everything you can to enhance working-memory skills. A double benefit derives from practicing working-memory exercises: the exercises prevent deterioration of the frontal lobes, which, in turn, increases your capacity to further enhance your working-memory skills.

Given its importance for optimal brain function, let me suggest *a series of exercises that can strengthen working memory:*

Gather a handful of pennies, nickels, dimes, and quarters and scatter them on a desk or table in front of you. You shouldn't count the number of coins ahead of time, but the desired number is anywhere between ten and fifteen of each, laid out in no particular order. Now pick them up one at a time and total the number of coins of each denomination. How did you do it?

Unless you have a highly developed working memory, you picked up one denomination at a time and totaled it before moving on to another denomination. It's unlikely you elected to pick up the coins at random while keeping a separate mental tally for each denomination and then totaling everything at the end. That's because it requires greater mental effort to keep track of all of the different denominations simultaneously. It's easier to count the coins one denomination at a time because that process makes fewer demands on working memory. After counting the nickels, for instance, that total can be stored in working memory and your attention shifted to the next denomination.

In order to increase your working memory, count the pennies and nickels at random (i.e., don't alternate them). This will require you to keep a running total of each denomination in working memory while you're counting. Do it rapidly and as you count each coin, place it to the side. When you're finished, write down your totals and then check for accuracy by separately counting each of the denominations among the discarded coins. Not much practice should be required for you to manage two denominations.

Next, count three denominations (pennies, nickels, dimes) in the same manner, and then count all four denominations. If you can manage four, you are achieving what is considered by psychologists to be the maximum. "Four items seems to be the limit. It's a fundamental characteristic of human working memory," according to Paul Verhaeghen, the psychologist at

Syracuse University who carried out the experiments demonstrating the four-item limit. (If you want to read the fascinating details of how he determined this, check his 2004 paper "A Working Memory Workout: How to Expand the Focus of Serial Attention from One to Four Items in 10 Hours or Less," published in *Journal of Experimental Psychology: Learning, Memory and Cognition*, vol. 30, no. 6.)

Let me suggest *several other exercises for enhancing working memory:* Over the span of one minute, name as many animals as you can. You can time yourself and make a tally mark for each word and count the tally marks after you've finished at the end of the minute. Also record your responses in order to detect duplications, since the name of each animal can be used only once. Better yet, ask someone else to make a tally mark for each word that you speak out. Twenty to thirty words is considered acceptable. Next, recite the number of cities you can think of in one minute. Then do fruits, vegetables, athletes, movie stars—whatever category allows you to choose from a vast number of potential choices in this test of what psychologists refer to as semantic (category) fluency. When finished with one category, switch to another while remaining within the one-minute time frame. For instance, alternate cities with fruits. Not all categories have been standardized; that is, there isn't an established upper or lower limit for movie stars. But aim for somewhere in the range of twenty to thirty items for each category within the one-minute limit.

Next, in a test of lexical (letter) fluency, recite within one minute as many words as you can starting with the letters *c*, followed by *f*, and then *n*. The words cannot be the names of people, places, or numbers, and they cannot be variations of the same word, e.g., *work, worked, working*.

Notice that both the category and the letter-generating fluency tests rely on language. They also activate the left frontal lobe. Anytime you want to "turn on" your left frontal lobe and enhance your working-memory abilities, you can do so by performing one of these exercises. One caveat: As you practice these exercises, you may begin to repeat earlier sequences. Counter this by restricting or shifting the categories (farm animals or jungle animals or animals found in rain forests or savannas). This exercises not only your working memory but also general memory.

The *second exercise for working memory* involves not words but designs. It activates an entirely different part of the brain: the right hemisphere, especially the right frontal and parietal lobes. Within four minutes draw as many novel (original) abstract designs as you can. The figures cannot be nameable (triangles, squares) and each one can appear only once. Page 124 shows a sample containing fifteen novel designs.

Here is another test of design fluency. Page 125 shows thirty-five squares with five dots arranged within each square. The placement of the dots within each square is identical. Photocopy the page so that you can take the test on multiple occasions. Now, using a copy, make within one minute a different design in each square by connecting the dots using only four straight lines. Each line must touch at least one other line at a dot. Five examples are drawn. It isn't necessary for the lines to include all the dots. But each arrangement

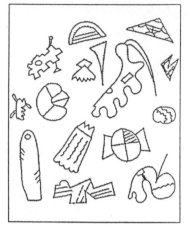

Example: healthy performance
Novelty Score = 15

Example: healthy performance
Novelty Score = 15

must be original—repetitions don't count, just as word repetitions didn't count in the word fluency tests. Initially you can expect to create between eight and ten original designs. With further practice you should be able to double that.

Now turn to page 126, which shows thirty-five squares, each containing five empty and five filled dots. Create original nonrepetitive designs using only four lines and alternating between empty dots and filled dots. You can start from either a filled or empty dot, but be sure to alternate them. Five examples are given. Try not to look at the previous designs as you make each new design. Check carefully at the end for repetitions or the use of five lines instead of four.

In order for you to challenge your nonverbal working memory, create on a blank sheet of paper with a ruler and a pencil a design of thirty-five empty squares arranged in a grid as on pages 125 and 126. By photocopying your

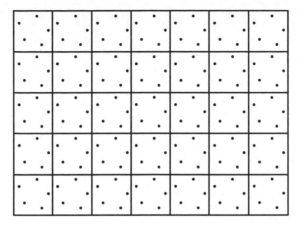

grid you can then create your own configurations of dot patterns.

The goal in each of the fluency tests is to activate the frontal executive circuits (primarily located in the dorsolateral prefrontal cortex). Each of the exercises described does

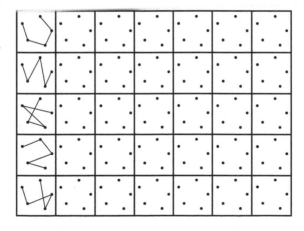

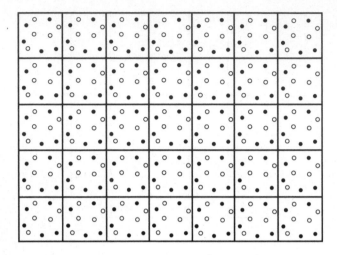

this. Just be sure to perform nonverbal design exercises as well as verbal exercises.

Look for opportunities to incorporate working-memory exercises into your everyday routine. For example, whenever

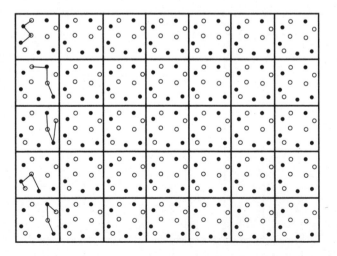

you're inclined to divide an activity into several segments, try doing them simultaneously as with the coin-counting exercise. When reading the paper, select three stories from page one and read the page-one content of all three of them before turning to the continuations on the back pages. That way you will be challenging your frontal lobes to keep two stories in working memory while you're reading one of the three. When you're finished, test your ability to remember specific names and facts from each of the stories.

Recreational games such as dominoes and bridge are exceptionally effective at boosting working memory. In one study, comparing fifty bridge players with nonplayers, all between the ages of fifty-five and ninety-one, the players consistently outperformed the nonplayers on tests of working memory and reasoning. Poker is probably just as effective, as is blackjack, especially when playing with a limited number of decks. If you really want to improve your working memory, learn to card count. This isn't overly difficult when only one deck is in use (the reason the blackjack dealers in casinos use multiple decks).

A highly recommended free website that offers programs for improving working memory is found at www.memorise .org/memoryGym.htm.

Expand Your Vocabulary

Although philosophers and most recently cognitive scientists continue to debate whether thoughts can exist without words and language, no one disagrees that words provide new and more nuanced ways of looking at everyday objects and experiences.

Learning new words not only enriches one's understanding of the world but also enhances several brain functions. Whenever you encounter a new word, you engage the language centers located principally in the left hemisphere; the word is then worked on by the prefrontal lobes and maintained in working memory. After continued study and thinking about it, the word becomes a component of your permanent vocabulary and is stored in long-term memory.

As part of your efforts to improve brain function, learn a new word each day. Keep a record of the new words you learn by entering them in a dedicated word journal. I started this exercise at twelve years of age in response to an intriguing and appealing proposal by my father. He told me that on certain days he would place a dollar inside the front flap of the dictionary. Since I couldn't be certain which days he would do this, he suggested that I check the dictionary every day and, dollar or no dollar, learn a new word each day. "Every day a new word will be your reward and on some days you will be doubly rewarded," he told me. Thus began a habit that has gone on now for almost half a century (sadly, minus the excitement of finding my father's dollar bribe).

When you settle on a word, you'll want to write down the word, its correct spelling and pronunciation, its various meanings, its derivation, any relevant history concerning the word, along with any personal reactions or observations about the word. Finally, you want to compose a sentence containing the new word. As an example, here is a brain-related word taken from Norman W. Schur's *2000 Most Challenging and Obscure Words:*

Cerebrate (SER uh brate) verb. To *cerebrate* is to use the mind, to think, or to think about. *Cerebration* (ser uh BRAY shuhn) is the working of the brain. *Unconscious cerebration* is the term used in psychology to describe the reaching of mental results without conscious thoughts. The American novelist Henry James (1843–1915) in *The American* wrote of "the deep well of *unconscious cerebration.* Both verb and noun are derived from Latin *cerebrum* (brain) which has been taken into English (SER uh brum; suh REE bruhm) to denote the front part of the brain, which controls voluntary movements and coordinates mental activity." In *The Human Comedy*, the American historian James Harvey Robinson (1863–1936) wrote: "Political campaigns are designedly made into emotional orgies which endeavor to distract attention from the real issues involved, and they actually paralyze what slight powers of cerebration man can normally muster." Note Robinson's date of death—many years before television served to exacerbate the situation.

Anyone can easily turn up new and intellectually stimulating facts about words with dictionaries and Internet sources. For instance, take a moment and do a quick Internet search by entering "Cerebrate definition." You'll learn about cerebralism, how to identify a cerebralist, and anticipate how he might respond to questions about the origin of the human mind. You'll learn how to recognize cerebricity when you encounter it. You'll also learn about cerebrates, powerful psychics (also referred to as "zerg cerebrates") from the StarCraft science fiction series of books and video games. These latter references can provide you with the material to compose your comments and responses to the new word.

After learning the word, mentally construct a sentence using it. As an alternative, try using the word in a sentence at some time during the day. Admittedly, this isn't always practical. Take my word for today: *scrutator*. "He answered the scrutator's questions" doesn't quite blend as smoothly into most conversations as "He answered the investigator's questions." In such instances, instead of using it in a sentence sometime during the day, simply incorporate the word into a written sentence.

Word exercises become even more challenging when you compose a short narrative incorporating as many words as possible from those you've learned during a week. This is easiest when the selected words share a common theme. One way of guaranteeing this is to subscribe (at no cost) to www.Wordsmith.org, a website that e-mails a new word each day to subscribers (informally referred to as linguaphiles: lovers of words). The site features a weekly theme that dictates the choice of words for that week. Using that site will make it easier for you to incorporate all or at least most of the words you've learned during the week into a short paragraph. Whenever you incorporate the newly learned words into a narrative, you'll be engaging those parts of the brain concerned with imagination—a vast and interconnected net-work that defies any attempt to establish firm boundaries.

After several months of collecting and writing new words in your journal, you'll be able to dip into the list at odd moments and mentally compose sentences and narratives containing as many of the words as you can manage. Finding and learning new words can be done anytime and anywhere a dictionary or computer is available. Best of all, everything that you've learned about the word will be entered into long-term memory. Each time you page through your word list, your memory for the listed words will be strengthened.

Spell More Challenging Words

In order to learn a new word, it's necessary to spell it cor-rectly. In an age of spell-checkers, finely honed spelling skills seem less important to some people than in earlier times. But there are more important reasons for enhancing your spelling prowess than simply limiting your dependence on not-always-available spell-checks. Spelling engages and stimu-lates several language-related brain areas and circuits. And spelling forces you to focus and mentally "see" the word prior to speaking it or writing it down. Finally, learning new words brings about brain changes according to the degree of diffi-culty involved in spelling the word: the greater the difficulty, the harder the brain has to work, as measured by greater brain activation.

If you've ever watched contestants competing in the

National Spelling Bee, you may have observed on their faces a look of effortful strain whenever they're asked to spell words in which the pronunciation provides little information about the correct spelling, such as *céilidh* (pronounced KAY lee), referring to a traditional Gaelic social dance in Ireland, Scotland, and Atlantic Canada. Words like that are difficult because the contestants must activate a different part of the brain in order to spell an irregular word correctly. With a regular word, for example *scan*, the sound corresponds closely to the arrangement of the letters. Words like *céilidh* or *yacht*, for which the sound does not closely match the spelling in English, activate areas of the brain that process word meaning, such as the frontal and parietal lobes, which process printed text. Regular words, in contrast, preferentially activate part of the superior temporal lobe devoted to the spelling of words in which the sound corresponds closely with the letters.

In Jeffrey Blitz's documentary movie *Spellbound,* a parent of one of the finalists in the National Spelling Bee offers this formula for correctly spelling either a regular or an irregular word: "Start by learning the meaning of the word, the language of origin, and the root of the word. Use the word in a sentence, repronounce the word, and review the meaning of the word. Then spell the word to yourself, match it with the sound that you're going to make when you speak it. After repronouncing the word one last time in your head, spell it slowly out loud."

Spellbound, as well as Myla Goldberg's 2001 novel *Bee Season,* have sparked intense interest in spelling bees. In 2006, 2007, and 2008, ABC provided live prime-time coverage of the final rounds of the Scripps National Spelling Bee. As a result of this keen interest in spelling, adults now want to be included in the fun and competition. The National Adult Spelling Bee is held yearly in Long Beach, California. The rules are similar to those of the Scripps Bee except for the age of the contestants. Anyone older than sixteen (the cutoff point for the Scripps Bee) is eligible to travel to Long Beach, pay the ten-dollar entry fee, and compete.

I was curious about the backgrounds of the winners of the 2006 and 2007 National Adult Spelling Bee, so I contacted them. Did their proficiency start in childhood, or later in life? While most spelling experts start nurturing their spelling expertise in childhood, Hal Prince, the 2007 winner of the National Adult Spelling Bee, wasn't especially interested in words or spelling until his early fifties. Since this pattern was so unusual, I asked Prince about his methods:

"First, I went through the dictionary recommended by the Bee page by page. I made a database of words for drilling and also made tapes of the words for listening while commuting or running. I borrowed or bought every book about words that I could find and went through them to find words that looked interesting."

I asked Prince if he attributes his success to a "gift" for spelling? "While I think I do have a facility for words and spelling, I suspect it's more like a top 10 percent rather than a top .01 percent. Mostly, it's just a matter of being interested in words and taking the time to study them."

Prince's intuition closely matches the findings of psychologist K. Anders Ericsson, who conducted a mail survey of the finalists of the 2006 National Spelling Bee: "The main find-

ing is that the amount of solitary study is by far the best predictor of success at the spelling bee competition," according
to Ericsson.

While you don't have to enter a spelling bee to get the
benefits of learning words and sharpening your spelling
skills, a bee provides structure and serves as a motivator. If
you're interested in joining or starting an adult bee, you can
contact Justin Rudd, the founder and organizer of the
National Adult Spelling Bee Championship, which is held
in May in Long Beach, California (JustinRudd@aol.com or
www.adultspellingbee.com). Before engaging at the highest
levels you might want to try your luck with a local spelling
bee. Local bees take place in coffee shops, bars, and performance centers across the country (they started in Brooklyn,
Seattle, and Portland, Oregon). In addition to individual
efforts, adult spelling bees also include team events, usually
three people to a team.

If you're shy or want to hone your abilities in private
before competing, Brandeis University maintains a great site
(www.Spellbee.org) where you can pit your spelling acumen
against an anonymous partner.

Whatever your preference, if you want to focus on spelling
as a brain enhancer, you should (thanks to the growing number of Internet resources and the increasing popularity of
adult bees) have no trouble finding an opportunity to compete with other enthusiasts for the competitive thrill of correctly spelling *propaedeutic, disembogue,* or *imbroglio.*

Get into the Habit
of Deliberate Practice

At this point, after engaging in exercises aimed at sharpening sense memory, general memory, working memory, vocabulary, and spelling, what does one do next in order to put everything together? In search of an answer to that question I spoke with K. Anders Ericsson, a bearded, trim, fiftysomething professor of psychology at Florida State University who has devoted his life to studying the basis for superior performance. Along the way Ericsson became convinced that most of us fail to achieve our more ambitious brain performance goals, not because we lack talent but because we fail to take advantage of our brain's capacity to respond when pushed to its limits.

While teaching at Carnegie Mellon University, Ericsson and his mentor, Bill Chase, studied the effects of practice on memory for numbers. Their first subject, a student identified simply as S.F., could memorize for later recall around seven digits—a typical phone number. Ericsson then methodically increased the span of numbers to be memorized. "When he got a digit list of a specific length correct, we would give him a list that was one digit longer," he said.

After practicing for an hour two or three times a week, S.F. was able to reproduce sequences of more than ten digits. With additional practice S.F.'s digit memory increased even more. After several hundred hours of practice, S.F. could remember without error sequences of more than eighty digits at a rate of one digit per second. In order to get a feeling for this, read the following string of digits and then close your eyes and try to recite as many of them as you can:

6513495272227500454680208713456553700678192 1
652344568075614503594923400960676596087

Unless you've previously engaged in digit-memorization exercises, you were able to recite probably about seven digits. This corresponds to S.F.'s best initial effort. But S.F. was able to increase his memory span for numbers by associating them whenever possible with running times for various track-and-field performances. An avid cross-country runner, S.F. encoded 358 as a very fast mile time, three minutes and fifty-eight seconds, just short of the four-minute mile. When the sequence included four digits starting with three (3493, for instance) he encoded it as three minutes 49.3 seconds.

"After deliberate practice, S.F., an average student, was able to recall a longer list of digits than anyone had ever done, even people with alleged photographic memory," Ericsson told me. "This result with S.F.—which we repeated with other volunteer students, incidentally—exemplifies to me the remarkable potential of 'ordinary' adults and their amazing capacity for change when they engage in deliberate practice."

As an illustration of what he means by "deliberate practice," Ericsson contrasts the practice methods of professional versus amateur golfers. "Most amateurs participate almost exclusively in recreational play with others. When they 'practice' they tend to do things they are comfortable with and can do with minimal control, such as whacking buckets of golf balls at a driving range. Professionals, in contrast, engage in practice activities that require full concentration to improve specific aspects of their performance. Further, they voluntarily choose practice routines in which they initially experience difficulties in order to improve a specific weakness."

Only by working on improving weaknesses can the expert golfer continue to enhance his or her performance. "The expert golfer's ability to perceive minute differences and exert control of the ball trajectories does not emerge naturally but through the process of acquiring refined mental representations for perceiving, monitoring, and controlling the muscles involved in the various required movements."

According to Ericsson: "For the superior performer in any field the goal isn't just repeating the same thing again and again, but achieving higher levels of control over every aspect of their performance. That's why they don't find practice boring. Each practice session they are working on doing something better than they did the last time. Intense solitary deliberate practice is the hallmark of the superior achiever in every competitive field that I have studied over my forty-year career."

Ericsson suggests two requirements for deliberate practice: full awareness and the avoidance of automated performances. As an example of a fully aware versus an automated performance, think back to when you first learned to drive a car. For the first few months you focused your full attention on learning the fundamentals. But as you became more proficient, your brain no longer needed to focus exclusively on driving and you could engage in other activities such as listening to the radio or talking to passengers. Now, except in emergencies and unusual situations, driving has become automated; you carry it out effortlessly and without conscious awareness of the various components of the driving experience.

Contrast your experience with the experience of a competition race-car driver. In order to become competitive, the driver must remain alert and intensely focused on innumerable variables. His goal is to push himself and his car to the outer limits of performance without risking personal injury or mechanical breakdown. Indeed, his success as a competitive driver will depend on his ability to continuously modify his driving performance in response to rapidly changing conditions.

The race-car driver differs from the everyday driver by virtue of ongoing vigilance and the monitoring of each component of the driving experience—in essence the deliberate practice espoused by K. Anders Ericsson. In contrast to the everyday driver for whom every driving experience is indistinguishable from every other, the competition driver can provide exquisitely detailed moment-by-moment descriptions of each race. Psychologists refer to this as a *flexible memory representation*. A similar combination of attentive awareness and flexible memory can be found among professional musicians like pianist Angela Hewitt, who writes of her own experience:

"In my recording sessions I find that the improvement comes not in endlessly repeating a piece, but in listening intently to what has been recorded and then thinking about how it can be done better. The editing process then becomes an art in itself and requires intelligent musical decisions."

In one experiment demonstrating the benefits of sustained practice, a group of skilled musicians memorized a piece of music and then, unexpectedly, were challenged to reproduce the same tempo under changed conditions: playing every other note, or playing notes with only one hand, or transposing the music into a different key. While experienced professional musicians had little difficulty meeting these unusual requests, less experienced musicians often couldn't do it. That's because the skilled musicians, as a result of their years of practice and performance experience, were able to meet the demands of a new situation by rapidly modifying their mental representation of the music.

Deliberate practice isn't "fun," doesn't involve doing what comes easily, and requires a high degree of motivation. "The requirement for concentration for improving performance sets deliberate practice apart from mindless routine," according to Ericsson. He points to expert violinists as another example of the power of deliberate practice, in this instance solitary practice aimed at mastering the specific goals determined by their music teacher at weekly sessions. "The greater amount of solitary music practice accumulated during development, the higher the levels of attained musical performance."

In time the intense concentration required for deliberate practice brings about alterations in the brains of extraordinary achievers. People with extraordinary abilities, it turns out, learn to use their brains differently from the average person. Take chess grand masters, for instance.

Measurements of brain activity during a chess game reveal that grand masters activate their frontal and parietal cortices, two brain areas known to be involved in long-term memory.

Skilled amateurs, in contrast, activate their medial temporal lobes, areas involved in coding new information. This preferential activation of the frontal cortex by the grand masters, who have memorized thousands of moves over their lifetime, suggests that they are using their long-term memory to recognize positions and problems, and to retrieve the solutions. Use of the medial temporal lobe by the amateurs, in contrast, suggests a less effective strategy of analyzing on a case-by-case basis the best response to moves and positions not previously encountered.

The grand master's expertise, in other words, involves storage in the frontal lobes of vast amounts of chess information. And this takes a long time and a lot of hard work. Grand masters typically spend a minimum of ten years amassing in their brains an estimated hundred thousand or more items of chess information (opening gambits, strategies, endgames). Thanks to this rich memory store, the grand master is able to quickly assess the advisability and potential consequences—many plays ahead—of a specific move. This ability is one of the reasons a grand master's performance against a good amateur always improves when moves must be made rapidly; there simply isn't time for a detailed and time consuming move-by-move analysis.

So, would learning more moves and deepening one's knowledge of chess turn an amateur player into a grand master? Not necessarily. The genius of the grand master depends not just on the amount of chess information stored in his long-term memory, but also on the organization of those memories and on how efficiently they can be retrieved. In essence, the grand master must put long-term memory to short-term use. Indeed, according to Brian Butterworth of the Institute for Cognitive Neuroscience in London, "experts develop a kind of long-term working memory [LTWM], which is specific to their area of interest."

An expert can be defined as a person who, after increasing his long-term memory as a result of many years of deliberate practice, incorporates that accumulated reservoir of information into LTWM. In essence, the more easily the expert can access information about his subject, the greater his expertise and control. And the more knowledge the expert brings to a subject, the more easily he can access that knowledge. After many years of deliberate practice and experience—ten years is considered the minimum—expert performance is attained.

There are many everyday examples of experts who are capable of increased temporary storage of information pertaining to their interests or occupation. An experienced waiter, rather than being limited to the usual seven items, can keep in his working memory the precise orders of up to twenty people without writing them down. An actor can recall his lines after reading them over just once. All of us can effortlessly understand and retain in working memory sentences of twenty words or more—well beyond our span for random sequential words (seven words) or words in a foreign language (three words).

Geniuses can store vast amounts of information in long-term memory and then retrieve it as circumstances demand. PET scan studies of geniuses and prodigies confirm increased frontal activity. For example, one study compared the PET scan of German math prodigy Rudiger Gamm with scans of people with no special calculating skill. When doing mental arithmetic, Gamm's brain, but not the brains of people who lack calculating skills, shows activity in frontal areas involved in long-term memory. It's speculated that Gamm uses his long-term memory to store the working results that he needs moments later to complete his mental calculations. Thus, when rapidly performing mental calculations he is less likely to lose his place. It's as if he were writing down on a notepad all the steps of his calculation so that he could later read them off. And if memory is indeed like a notepad—an analogy I borrowed from scientists who study calculating geniuses—then the memory of a Rudiger Gamm is like a library of notepads containing within his LTWM for instant retrieval every calculation he has ever carried out.

In 2005, I personally confirmed the power of deliberate practice. I was in La Jolla, California, to deliver a lecture on music and the brain immediately before a performance by concert pianist Gustavo Romero at the Mozart Festival. After the concert, which featured several of Mozart's piano sonatas, Romero invited me to accompany him to a dinner party in his honor. At one point in our conversation that evening, I asked Romero how much time he spends practicing each day. He answered: "With the exception of those times when I'm in transit from one concert to another, I practice between four and six hours a day. When I get to my concert destination, I make up for lost time by increasing the hours of practice over the next few days. I've kept to this schedule since starting to play the piano at age eight."

As with the grand masters, Romero has stored within his frontal lobes all his special knowledge, in this case the different compositions of Mozart. So is he a genius who has simply coincidentally logged many thousands of hours of practice? Or does his genius consist of his willingness to put in those hours?

In other words, is the ability to form larger-than-normal long-term memories a genetic trait? Or does it depend on one's individual effort? Important implications ensue from the answer to this question.

If genius is entirely genetic, then the average person's reservations about achieving success at the highest levels would seem to be justified. Indeed, if genes are more important than environment, most people are likely to remain permanently mired in mediocrity. But if, on the other hand—as suggested by the research of K. Anders Ericsson on deliberate practice and by the Flynn effect of intelligence researcher James Flynn—individual effort can lead to enhancement of the brain's structure and function, then most people may be capable of achieving levels of performance that will separate them from the vast majority of their competitors. Indeed, superior performers sometimes admit when pressed on the subject that their achievements depend more on hard work than genius or genetic endowment.

In a famous interview with Larry King, Marlon Brando stated that with proper training anyone could be an actor. The novelist Graham Greene made a similar point when asked about his writing talent: "One has no talent. I have no talent. It's just a question of working, of being willing to put in the time."

I put this question of inheritance versus development secondary to increased effort to K. Anders Ericsson. He is on the side of Brando and Greene, firmly convinced that special inherited qualities aren't what distinguish people with expert abilities from people of more humble accomplishment. The key ingredient turns out to be the willingness to "stretch yourself to the limit and increase your control over your performance," he said.

For proof, Ericsson points to a study he carried out at the highly regarded Music Academy of West Berlin. "Superior" students, judged by their teachers as most likely to go on to concert careers, put in an average of twenty-four practice hours per week. "Good" students, thought more likely to end up as teachers than performers, practiced an average of only nine practice hours per week. Ericsson found a similar pattern of intense solitary deliberate practice among superior performing athletes, chess players, mathematicians, and memory virtuosos.

A 2007 study of chess players carried out at Oxford University confirms Ericsson's emphasis on the power of deliberate practice. The researchers studied fifty-seven primary and secondary school chess players, logging their hours of daily chess practice as well as challenging them with chess problems and IQ tests. Although IQ and years of experience contributed to chess ability in the fifty-seven students, the highest correlation was with the number of hours a day the participants in the study played chess or studied classic chess games. Among the top twenty-three players the correlation of chess ability with IQ disappeared altogether. And within the group with the highest IQ, playing ability aligned with the number of hours practiced rather than the IQ score.

Ericsson's claim that exceptional memorizers are "made, not born" was confirmed in 2003 when ten of the world's foremost memory performers, drawn from the World Memory Championships, were pitted against ten people with ordinary memories but equal intelligence and spatial abilities. As a first step, both the memory performers and the controls submitted to structural MRI images of their brains. No differences were found. Brain activity was then recorded while members of both groups memorized digits, black-and-white photographs of faces, or snow crystal patterns.

Not surprisingly, the memory superstars greatly outperformed the members of the control group on digits. But no differences could be detected on memorizing snow crystal patterns (unusual and difficult to verbalize). On facial memorization the memory performers remembered more faces, but the differences between the two groups weren't striking.

Next, all of the participants were quizzed about their memorization techniques. Nine of the memory performers but none of the controls used the mnemonic strategy of mentally placing the items to be memorized within an imagined place, such as within one's living room. (This memorization "method of loci," dating back to 477 B.C., is described in Part Three.)

In the final part of the experiment everyone in the study underwent an fMRI while memorizing. One dramatic difference separated the two groups: those with superior memories engaged brain regions that are critical for spatial memory.

"In essence, we found that superior memory is not driven by exceptional intellectual ability or structural brain differences," concludes Eleanor Maguire, the principal investigator in the study. "Rather, superior memorizers use a spatial learning strategy and engage brain regions that are critical for spatial memory."

Maguire's findings provide "compelling evidence," according to Ericsson, that "ordinary people can dramatically improve their performances in various areas by means of deliberate practice and the development of novel strategies."

History provides additional support in favor of individual effort rather than genes. Take athletics, for instance. In marathon and swimming events, many serious amateurs active today could easily outperform the gold medal winners of the Olympic Games of the early twentieth century. Further, some performances that are now considered routine (double somersault dives, for example) were once thought impossible or too dangerous.

After the fourth Olympic Games of the modern era, in 1908, the Olympic Committee prohibited double somersault dives because they were considered impossible to control. Today, as a result of deliberate practice, competitive divers have moved well beyond double somersaults and routinely perform dives of far greater complexity. Runners Roger Bannister and John Landy provide another example. When they both broke the four-minute mile in 1954, their performances were considered extraordinary. Today the ability to run a mile in less than four minutes isn't going to guarantee a victory at the higher levels of competition. Such examples from competitive sports illustrate that, with the exception of height and body size, our bodies and nervous systems are modifiable.

In practical terms, research confirms that exceptional performers aren't endowed with "superior" brains. Rather, the brain, thanks to its plasticity, can be modified by deliberate practice and the use of innovative strategies. That combination will enable you to achieve high levels of performance in the area of your choice—*if you are willing to put in the effort required to achieve mastery.*

PART FOUR

Using Technology to Achieve a More Powerful Brain

I f you're over forty, you may not be terribly familiar with video games. But you need to do something about that if you're serious about improving your brain's performance. That's because when wisely used, video games, especially action-video games, can help you notice more, concentrate better, respond more quickly, and increase several components of your overall IQ.

Basically, properly designed video games induce a more efficient pattern of activity in action-related brain areas. They do this by reorganizing brain activity so as to increase perceptual and motor speed as well as the increased eye-hand coordination needed for video games.

The effects induced by regular video-gaming can be compared to what occurs in the brain of a concert pianist. As a result of many hours of practice, the pianist's brain has built up more efficient networks of neurons in the supplementary motor cortex, the brain area responsible for planning the finger movements required for each selection. With increased experience and practice, fewer and more efficient networks are recruited for the performance of each musical selection.

The principal benefit from video-game simulation is the acquisition of highly specific real-world skills. Probably the most familiar examples of this are the air-flight and air-traffic-controller simulators. Pilots and air traffic controllers routinely spend hundreds of hours on simulators in order to hone their skills. Unfortunately, their increased skill in handling flight-related situations doesn't transfer to other life situations: if you're scanning a crowd in search for a friend, you're not going to find him any more quickly by enlisting the aid of an air traffic controller or pilot. That's because their video training is highly specific. As a result, when it comes to attention *in general,* the controllers and the pilots aren't any better than the rest of us.

But you might find your friend in that crowd more quickly if you ask an avid action-video gamer for assistance, based on the research of Daphne Bavelier, associate professor of brain and cognitive sciences and a member of the prestigious Center for Visual Science at the University of Rochester.

Bavelier discovered enhanced visual search skills in action-video gamers during an ongoing study carried out with one of her students, C. Shawn Green (himself an avid gamer). They compared students who played action-based video games with students who didn't play video games at all (an under-represented group on the average campus, as Bavalier and Green discovered). The video-playing contingent were not just casual players but were "hard-core" gamers who had played action-video games like Counter-Strike at least three days a week over the previous six months.

In the experiment all of the participants in the Bavelier-Green study rest their heads on a chin rest and stare at a square in the center of a computer screen. Randomly a target (a circle with a triangle enclosed within it) flashes at one of twenty-four possible locations on the screen. Immediately the screen is flooded for about a second with a clutter of circles, squares, and lines. Finally, the screen goes blank and the participants are asked to remember where the target had originally appeared on the screen. Regular video-game players do this with 80 percent accuracy, while nonplayers get it right only about 30 percent of the time.

Perhaps you're thinking, "These differences could be the result of self-selection: people with superior visual attention might naturally be attracted to video games." To cover that possibility, Bavalier and Green scrounged around the campus and after a good deal of effort came up with thirty-two non-video-playing students. Half of them were assigned to play the puzzle game Tetris, and half to play the action game Unreal Tournament. All played their assigned game for thirty hours over the course of about thirty days.

In case you're not familiar with Unreal Tournament, it features a cornucopia of enemies and hazards coming at you on the screen from every direction. Allow your attention to waver for a few milliseconds and . . . lights out! In contrast, the classic falling-blocks puzzle game Tetris is far less visually demanding. Since the puzzle pieces always drop from the top of the screen, there isn't any need to scan elsewhere.

At the end of the study, those students playing Unreal Tournament showed a 15 percent to 20 percent improvement in their ability to ignore visual distractions. The Tetris players showed no improvement.

With practice, "video-game players can process visual information more quickly and can track more objects on a computer screen than nonplayers," according to Bavelier. Two processes serve as limiting factors in tracking objects on a screen.

The first, *attentional* blink, is the half-second recovery time required to detect a second target during a rapid-fire sequence of targets. For example, if I ask you to watch for a white letter appearing on a computer screen at some point in a stream of black letters, you'll have no problem doing so. But if I then ask you to look for an X appearing after the white letter in a position ranging from immediately after the white letter to eight letters later, things become more complicated.

No problem if the X comes three or more letters later. But if the X appears very close in time to the white letter, you'll probably miss it. That's because while your brain is "busy" looking for the white letter, it will be unable, because of attentional blink, to process anything else. This brief but critical gap can be narrowed if a person spends several hours a week on a regular basis playing action-video games. In essence, action-video games enable the game players to shorten their attentional blink and thereby perceive and respond to threatening targets that typically spring unexpectedly from the periphery of the screen.

The second process, *subitizing*, refers to the ability to look at an array of objects and immediately and correctly enumerate them without resorting to counting. For example, when you quickly glance at the checkout lines in the supermarket and without counting automatically select the shortest line, you're subitizing. Most people can do that with up to four objects. Anything beyond that subitizing limit of four must be counted and requires extra time, 250 to 350 milliseconds, for each additional item. But for action-video players, the subitizing number is 50 percent improved. What's more, only one hour a day for ten days of action-video gaming with an emotionally intense shooting game such as Medal of Honor is sufficient to improve both visual attention and processing time. No improvement in either factor occurs among players of Tetris or other non-action-video games, Bavelier and Greene discovered.

Although no one has so far come up with a completely satisfying explanation for these differences, I suspect they're due to the increased threat and fear levels experienced by players of games like Medal of Honor, where players can lose their "lives" rather than simply fail to solve a puzzle as with Tetris.

This arousal of fear and aggression in response to perceived threats also plays a part in explaining why violent video games incite violent behavior in certain predisposed players. The amygdala, a small almond-shaped nucleus below the cerebral cortex that responds to threatening or fearful faces, doesn't distinguish between events occurring in a game and the same thing happening in "real life."

"Playing action-video games can alter fundamental characteristics of the visual system," Bavelier says. Essentially, video games provide a convenient and easily accessible means for changing the brain. The resulting fine-tuning of visual attention will enable you to see more, and respond more quickly and more accurately to simultaneously occurring events.

Cognitive Versus Physical Fidelity

In order to understand the value of digital game-based learning as a training tool for brain enhancement, it's helpful to distinguish between what's called cognitive and physical fidelity. Physical fidelity means that the training program faithfully replicates the real-life situation. For example, the early versions of the air-flight simulators consisted of the front of a real airplane hooked up to computer displays. Sitting in one, you almost couldn't tell whether you were in the cockpit of an actual plane or in a simulator.

Several years ago I experienced firsthand the effects of digital virtual learning–flight simulation based on physical fidelity. An airline pilot patient of mine was required by the FAA to undergo simulator testing after recovering from a recent head injury. As part of the evaluation, the FAA requested that I accompany him to the testing so that I could be interviewed about the risk that he might experience an epileptic seizure as a result of the head injury. While in the simulator my patient performed so well that he successfully completed the testing a half-hour early. Since a half-hour of paid-for time remained, the evaluators asked if I would like to try the simulator. Caught off guard, I readily agreed to what I anticipated would be an experience no different from playing a video game.

A few minutes later, I found myself strapped into a seat and at the controls of what had once been the cockpit of a real 747 but was now part of a simulated testing protocol. I remember two things about the experience: the initial thrill of looking through the windshield and observing how the scene changed as I moved the yoke (control wheel); and the uneasiness I immediately experienced when the instructor told me after several minutes of admittedly enjoyable flight simulation, "It's now time to land the airplane. Listen to my instructions and I'll tell you how to do it." Of course I reminded myself for the umpteenth time that the simulated flight wasn't "real." But somehow that didn't calm me down.

My fear further escalated when the "plane" rattled and shook violently as I landed it clumsily on the tarmac. At this point my hands started shaking; they were still shaking five minutes later when over iced tea the instructor explained what had gone wrong. Apparently at the last second I had been so rattled that I didn't process his instructions correctly and pointed the nose of the plane down instead of flaring it up, resulting in a jolting, bone-jarring landing that in a real-life situation would have risked shearing off the nose wheel, causing the plane to spin totally out of control.

Looking back on the experience, I realize now that my conflict and apprehension resulted from my brain's attempt to reconcile two conflicting interpretations of what was going on. I realized that the flight wasn't "real" and that I couldn't be harmed. Yet the combination of an actual airplane cockpit, the video imagery that changed in response to my movements of the control, and the shaking and rattling that coincided with my inexpert landing led to the powerful illusion that somehow I was *really* piloting an airplane.

During the debriefing, the evaluator told me—in an effort, I suspect, to make me feel better—that even some veteran pilots could be made to experience similar anxiety. "When we started using simulators, we used to push everybody to his or her limit," he said. "Essentially, we kept adding challenges like the loss of an engine on takeoff, being hit by lightning, a sudden loss of altitude on final approach due to wind shear, loss of hydraulic pressure making it impossible to steer the aircraft, or one of the landing gears failing to retract. We kept pushing the pilots until eventually the plane crashed. The pilot who was the last to go down in flames earned the highest rating. We don't do that anymore. We found that crashing a plane—even a virtual one—can set off tremendous anxiety in a pilot that sometimes takes a lot of debriefing to overcome."

Today the emphasis in training programs has shifted from physical to *cognitive fidelity*. In contrast to physical fidelity, which uses some components of the real situation, cognitive fidelity faithfully incorporates only the relevant mental processing.

For example, the program "Space Fortress" is an attention enhancer aimed at improving pilot performance even though the game isn't specifically related to pilot training. There are no simulated airplane cockpits or anything else to suggest that the test has anything to do with flying an airplane. "The key to success in training is to ensure that the cognitive demands in training resemble those of the real-life task," suggests Professor Daniel Gopher, one of the world's leaders in the field of cognitive training and author of the pioneering paper "Transfer of Skill from a Computer Game Trainer to Flight."

"Ten hours of training with 'Space Fortress' resulted in 30 percent improvement in flight cadet trainee performance. When we compared 'Space Fortress' with a sophisticated, pictorial, high-level graphic and physical fidelity-based computer simulation of a Blackhawk helicopter, 'Space Fortress' proved successful in improving performance while the other, a NASA-sponsored program, did not."

Nor are the benefits of cognitive fidelity training programs limited to airline pilots. Video games (digital game–based learning) are currently being used for everything from determining sources of water contamination (MIT's Environmental Detectives) to learning German to training marines (MAK's MAGTF-XXI) to dealing with a bioterrorist attack (Carnegie Mellon's Biohazard/Hazmat) to treating neuropsychiatric conditions such as fear of heights, social phobias, addictions, and post-traumatic stress disorder (PTSD) in veterans from the Iraq war. As an indication of the verisimilitude of action-video games, the PTSD simulation is a modification of the Xbox game Full Spectrum Warrior.

"The inherent brain mechanisms for performing complex skills is different in highly experienced video gamers," says Joshua A. Granek of the Centre for Vision Research at York University in Toronto. "Video-game training reorganizes the brain's activity and leads to more efficient and effective control of skilled movements other than playing video games."

In support of Granek's claim, surgeons who play video games more than three hours per week commit 37 percent fewer errors in the operating room, are 27 percent faster at laparoscopic skills (surgery involving the maneuvering of instruments on the basis of images transmitted from a tiny camera placed at the operative site), and are 33 percent faster at suturing than surgeons who don't play video games.

Video Games and the Mature Brain

In older adults some astounding brain enhancements have been documented via the use of video games. In one of the earliest studies, elderly subjects (ranging in age from seventy-one to seventy-eight years) improved their scores on tests for both verbal and nonverbal intelligence after only one hour of video-gaming per week. As a result of such positive reports, older players are currently the fastest-growing segment of video gamers. According to an internal customer survey at PopCap Games in Seattle, more than 70 percent of its players are older than forty, with almost half of them older than fifty. What's more, older players tend to spend more time playing per session than their younger counterparts. In one popular free site (Pogo.com), players fifty years of age and older accounted for more than 40 percent of the total time spent on the site.

While no one has convincingly demonstrated that video-game playing can prevent or even delay the onset of dementia, there is general agreement that short-term memory and speed of response times are enhanced among video gamers. This decrease in reaction time is especially intriguing and raises an interesting and so far unanswered question: Does IQ increase in tandem with reaction time improvements?

Since many tests of intelligence emphasize faster processing speed, IQ strongly correlates with performance, at least on these tests. In general, people who are quickest on reaction-time tests tend to have above-average IQs. But other factors may also influence reaction time: some people are too cautious, lack self-confidence, or become easily distracted. Nonetheless, even though the correlation between speed of processing and IQ isn't completely proven, playing video games will increase your ability to make speedier choices, whatever their effect on your IQ.

"The results from these studies suggest that video game experience could be a powerful tool in slowing, stopping, or even reversing the age-related declines in perceptual, motor, and cognitive capabilities that regularly occur in the elderly population," according to Shawn Greene.

Obviously, video games aren't the only source of cognitive enhancement in the mature brain. Similar benefits in combating age-related decline can be gained from other cognitively challenging pursuits such as bridge, chess, and puzzles. None of these can be counted on, however, to increase processing speed. Video games are unique in this regard.

Do I have any caveats about taking up video games? Just two:

First, some preliminary evidence indicates that video games can be habit-forming, even addictive. This is especially true for those games that allow for high-intensity *immersion*, the tendency for players to become totally absorbed in the video game: skipping meals, staying home from work, ducking family obligations. Especially worrisome is what has come to be called *situated immersion*, not just playing the game, but actually experiencing the illusion of existing within the game world. Situated immersion increases as video games advance in the intensity, vividness, and quality of the video graphics; the level of control the player can achieve within the game; and the extent to which the game offers an intriguing alternative to everyday life. All three of these elements are currently increasing at a blistering pace with the newest games providing near total immersion. This may be especially perilous for people who experience anxiety around other people—sufferers from so-called social phobia.

"Games offer virtual worlds that are often more exciting and beautiful than real life," says psychotherapist Shauvan Scott. "We can achieve a sense of power and success that we may not be able to achieve in real life." Scott, a former "compulsive gamer," warns that "the dependency on video games can be debilitating."

To avoid the perils of immersion, limit the amount of time you spend playing video games to no more than two to three hours per week. Also limit single sessions to an hour maximum. If you spend more time video-gaming, you increase your chances of inducing some of the symptoms of attention deficit/hyperactivity disorder, especially such hyperactivity symptoms as a subjective feeling of restlessness accompanied by a need to be "on the go." This is more likely to occur with some of the more frenetic games.

What effect will video games have on attention? The prevailing opinion among the experts I polled doesn't suggest that video games worsen a player's attentional powers—the games enhance concentration and focus, at least while playing the video game. To monitor your own power of sustained attention, try a variation of what I call "the Henry James test." After a video-game session, "power down" by reading a chapter of *Portrait of a Lady* or *The Golden Bowl*. If you find yourself feeling restless and jumping ahead in the story in anticipation of James "getting to the point," you've spent too much time on the video game.

The second caveat (a more controversial one): Avoid games like Manhunt, Grand Theft Auto, and Take-Two, which feature gratuitous violence. Avoid as well games based on actual large-scale rampages or massacres such as Columbine or Virginia Tech. Although it's difficult to prove that such video games "cause" people to act violently in real life, the evidence is pretty good that they can engender unhealthy mental states and, in some people, antisocial personality trends. Certainly they can create anxiety and other uncomfortable emotional states in many players (speaking from my observations and discussions with other players as well as my personal experience). Why does this occur? What is the mechanism within the brain that explains it?

Part of the explanation has to do with the realism of the display. For instance, watching a video that displays realistic human hands in a realistic environment activates the motor areas in the viewer's brain. Geometrical or cartoon versions of the hands don't. The depiction of movement is also important. "Even when the superficial resemblance to a human is slight (aliens), if the agent moves like a human, it is more likely to be interpreted as being humanlike," writes India Morrison, an expert on video-game players' empathy with video-game characters. This selective response pattern is likely related to activation of the brain's *mirror neurons*, located in the frontal areas of the brain.

Mirror neurons first came to attention only a few years ago based on findings in macaque monkeys. Researchers observed that a cluster of cells in the prefrontal cortex of the monkeys fired both when a monkey grasped a peanut and when it watched another monkey grasp it. Only the sight of another monkey performing the action of grasping activated the mirror neurons. Watching a robot or mechanical device grasp the peanut failed to activate the mirror neurons.

Similar mirror neuron activations occur in our own brains when we watch the actions of another person. Some of these activations are highly individualized and based on life experiences that have sculpted the brain in specific ways. Pianists, for instance, show mirror neuron activation while watching someone play the piano; no activation occurs in the brain of someone lacking piano training.

When it comes to everyday experiences, mirror neuron activation patterns are similar from one person to another. Watching someone eat or drink, for instance, activates similar neurons in the observer that would be activated if she was actually eating or drinking. At the most basic level, mirror neurons provide the mechanism by which one person identifies with another.

Does such identification with another person extend to video characters too? Good reasons exist for believing that it does. In real life, emotional as well as motor behavior tends to be "mirrored" during our social interactions with other people. This helps explain why most of us tend to avoid people who consistently express negative or distressing emotions; after a few minutes in their company we begin to feel similar negative emotions ourselves on the basis of our brain's reading (at an unconscious level) of their facial expressions, tone of voice, and bodily movements. Psychologists refer to this process as *visuoaffective mapping*. Facial expressions denoting disgust, pain, or fear are especially effective in arousing similar emotional states in us when we observe those facial expressions. Both the facial expression and the accompanying emotions are mirrored and unconsciously we adopt them.

A similar situation exists in the world of video games such as Second Life, where avatars (three-dimensional-looking models serving as players' self-representations in an Internet world) have become more lifelike over the past decade, thanks to advances in computer graphics. Skin, hair, and fabric are now rendered in near-perfect verisimilitude. But even more important advances have taken place in the rendering of an avatar's social expressions: facial configuration, eye-gaze direction, gait. As a result, users are now more likely to infer character traits from an avatar's behavior and appearance.

With further refinements, this attribution of human emotions to video simulations is likely to increase. "It is quite conceivable that in a few years avatars will be available whose behavior is nearly imperceptible from humans," writes Judith Donath of the Massachusetts Institute of Technology Media Lab.

A correlation between video games and emotional experience is of special concern because, as Mike Musgrove of *The Washington Post* puts it, "consumer electronics makers looking to tap into the growing computer and video game market are seeking new and novel ways to help players feel as if they're inside their favorite virtual world." (A good reason to consider carefully what kind of virtual world you want to be in.)

But despite their potential for mischief, I believe video games have a lot to offer. As new and more sophisticated games come on the market, additional brain enhancement capabilities will become widely available. "The surge in new video games being developed to enhance one particular trait or another is probably the best testimony of the level of excitement in this field," says Shawn Greene. "Video-game research is opening a fascinating window into the amazing capability of the brain and behavior to be reshaped by experience."

In summary, if you take up video-gaming, even casually, you can expect the following benefits: decreased overall reaction times, increased eye-hand coordination, and enhanced manual dexterity. You'll improve your spatial visualization skills and your ability to mentally work in three dimensions. Finally, you'll be better able to divide and rapidly switch your attention as well as increase the number of things that you can visually attend to simultaneously.

Brain Gyms

Internet-based "Brain Gyms" are another increasingly popular technological approach to brain enhancement. These computer-based programs serve as a kind of "Grandpa Einstein" that, it's claimed, can make up and even reverse age-associated declines in memory and mental fitness in general. So far, though, the value of these programs hasn't been established, since a really solid study, such as the Bavelier and Greene studies on the cognitive benefits of action-video games, hasn't been carried out. Such an evaluation is hampered by the wide variations among the different programs:

some of them feature listening exercises, while others take a more visual approach; some involve a good deal of physical activity, while others require nothing more than moving a computer mouse. Nonetheless, one company, the producer of MindFit, claims on the basis of an internal comparison study that regular use of its product leads to greater gains in short-term memory and attention than those occurring among players of video games.

Having personally tried several of the Brain Gym programs, I think that action-video games are more likely to lead to brain enhancement. For one thing, video games are more entertaining than the repetitive "mental workouts" provided by the various brain fitness programs. Tracking balls around the computer screen doesn't begin to capture one's interest as successfully as playing one of the more sophisticated action games. The situation is like comparing a workout on the treadmill at the gym to running the same distance along a beach.

Brain Gyms may improve substantially in the future, especially if they incorporate some of the features of action-video games. But at the moment, the best way of deciding if a Brain Gym approach is for you is to personally try out one of the programs. I suggest starting with a free one or one that offers a limited-time free trial offer. After a few sessions, switch to an action-video game and compare the two experiences. I think you'll find that memory and attention are more easily enhanced through a challenging video game than through the mental exercises found in Brain Gyms, which after a while become uncomfortably reminiscent of the classroom.

PART FIVE

Fashioning the Creative Brain

I f you employ the methods suggested in the past chapters, you will notice improvements in your brain's performance. Your general memory and working memory will be strengthened, your attention span increased, your spelling and vocabulary improved, and your visualization skills enhanced. Taken together, these improvements should lead to greater creativity.

As a prelude to discussing the brain and creativity, it's helpful to say a few words about problem solving in general. Usually we solve a problem via a series of distinct steps. Consider the problem on page 172, "How many dots will be in the next figure in the sequence?" In order to answer that question, we first familiarize ourselves with the elements of the problem by reading and thinking about it. Our attempt to achieve a workable solution activates various and sundry brain circuits. For example, in this problem we notice that the first figure consists of twelve dots; the second figure twenty dots; and the third figure twenty-eight dots. In the progress from figure one to figure three, each figure is increased by eight dots, suggesting that the fourth figure might have thirty-six dots. That seems too easy—there must be a trick here, but what is it?

At the second stage of problem solving we look for less obvious solutions—a process that involves the activation of

DOT MATRIX
How many dots will be in the next
figure in the sequence below?

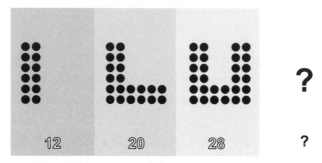

Hint: *Visualize the final figure and count the missing dots.*

additional brain circuits. The key element here is to pay attention, focus on the problem, and inhibit irrelevant brain processes. If the solution still isn't forthcoming, then we can further arouse the brain by increasing its oxygen supply (deep breathing; a short spurt of exercise such as push-ups or a run around the block), or we can shift the brain's activation patterns away from the circuits associated with the words and logic by listening to instrumental music.

Finally, if a solution still isn't forthcoming we can put the problem "out of mind" for a short period. Rest, relaxation, sleep—especially dream sleep—can help spawn the innovative associations needed for creative problem solving. (The answer to this puzzle is on page 194.)

Use Sleep to Boost Creativity

Not only do we remember more of what we've learned after "sleeping on it," we're also more creative, according to Jessica Payne, an upbeat young sleep researcher from Harvard University. During my conversation with her, Payne told me of her experiment proving this. It involved volunteers learning a list of related words (for example, *thread, pin, eye, sewing, sharp, point, haystack*). After learning the list, Payne's subjects were asked twelve hours later to recall as many words as they could. It turned out that their performance depended on whether or not they had slept in the interval. Half of the volunteers learned the list at nine a.m. and after a normal day for them returned for testing at nine p.m. The other half learned the list at nine p.m. and returned for testing the next morning at nine a.m.

Not only did the sleep group perform better in remembering the list, but those who had slept between the two testing sessions recalled two to three times as many novel or creative words.

"We found that sleep enhances creative associations between the words on the memorized list, thus encouraging the discovery of new and meaningful connections," Payne enthusiastically exclaimed.

As Payne described her research, I thought of a famous experiment carried out many years ago by sleep researcher William Dement. He showed his students the letter sequence H, I, J, K, L, M, N, O and asked them to name the one word suggested by this sequence. They had to come up with a specific word relating to this particular string of letters. Responses such as "letter sequence" or "alphabet" weren't allowed.

When none of the students came up with the correct reply, Dement suggested they "sleep on it" and see if the solution occurred to them. Next morning several of the students returned and reported their dreams.

One student reported several dreams. In one dream he encountered a barracuda while skin diving; in another, he was hunting sharks; in a third, he was sailing into the wind. Given those dreams, it can be reasonably concluded that the dreamer knew the answer on a subconscious level and had represented it repeatedly in his dream images even though he couldn't come up with the answer: *water* (the sequence of letters proceeds from H to O, i.e., H_2O). The student had correctly arrived at the solution to the problem via his dream images in which water was the common element. None of the students who had not "slept" on the puzzle came up with similarly suggestive images.

Although dreams played a prominent part in the student's solution to Dement's puzzle, dreams aren't necessary, according to Payne's findings. Sleep alone enhances memory and consolidates it. This provides two distinct benefits: we remember more, and, as suggested in her experiments, our memories are more likely to be creative. "We found that sleep changes memories in a manner that encourages the discovery of new and meaningful connections," Payne commented.

In his book *This Year You Write Your Novel*, Walter Mosley has this to say about how sleep furthers creativity: "While you sleep, mountains are moving deep within your psyche. When you wake up and return to the book, you will be amazed by the realization that you are further along than when you left off yesterday."

REM sleep is especially effective in enhancing creativity. As mentioned earlier, REM is the stage of sleep when dreams occur. During REM weak mental associations are strengthened and flexible creative processing enhanced. The most famous example of sleep-associated creativity is August Kekulé's discovery of the structure of the benzene ring. The chemical arrangement of benzene's six carbon and six hydrogen rings came to Kekulé in a dream.

As a student Kekulé had testified before the equivalent of a grand jury investigating the sudden and unexplained death of one of his neighbors. Among the deceased's belongings was a gold ring fashioned in the shape of two intertwined snakes biting their own tails—the alchemical symbol of the unity and variability of matter. After completing his testimony Kekulé never again consciously thought of the snakes.

One night many years later, after working unsuccessfully on the chemical arrangement of benzene's six carbon and six hydrogen atoms, Kekulé "turned [his] chair and sank into a doze." He dreamed of atoms swirling and dancing, joining together and then pulling apart in a snakelike pattern of motion. At that moment, in Kekulé's words, "one of the serpents caught its own tail and the ring thus formed whirled exasperatingly before my eyes. I awoke as by lightning and spent the rest of the night working out the logical consequences of the hypothesis."

Upon awakening, Kekulé envisioned the benzene molecule as a six-layered structure composed of carbon atoms with hydrogen atoms suspended like charms from a bracelet. The unconscious memory of the signet ring revisited in his dream had stimulated the creative visualization needed to discover benzene's structure.

Recent advances in neuroscience suggest what's happening in the creative brain during each of the "power down" states (rest, relaxation, sleep, especially dream sleep). They all share a common mechanism: a change in the brain's neurotransmitter systems.

"Creative individuals have a special ability to modulate the norepinephrine system, especially the connections of that system with the frontal lobes. During creative innovation, cerebral levels of norepinephrine diminish, leading to the discovery of novel orderly relationships," says Kenneth Heilman, a distinguished professor of neurology at the University of Florida College of Medicine in Gainesville.

A scholarly-appearing man with thin steel-rim glasses and a penchant for bow ties, Heilman has thought a lot about creativity during his own highly creative career of over half a century as one of the nation's premier behavioral neurologists. During our conversation on creativity Heilman emphasized the importance of the frontal lobes and how we can enhance their functioning.

"The frontal lobes appear to be the part of the cortex that is most important for creativity. They are critical for divergent thinking—the ability to diverge from what one has been taught to believe—and the associated ability to come up with alternative solutions."

To illustrate Heilman's point, look again at the dot-sequence puzzle. Concentrating just on the dots leads to the wrong answer (no, thirty-six is not correct). So what other way can we approach the problem? Hint: Instead of thinking strictly in terms of dots, visualize the *geometry* of the evolving figure. With the two additions of eight dots ($12 + 8 = 20 + 8 = 28$), the figure begins to take on the conformation of a square. But it's not quite a square. So how many dots would be required to complete the square? The correct solution (the addition of four dots to make a total of thirty-two dots) only comes to mind when you shift the emphasis from the dots to the figure as a whole. Only by engaging in such a shift in your thinking can you solve the puzzle. You must redirect your attention from counting dots to thinking in terms of a geometric figure (a square) formed by the dots. Divergent thinking provides the key insight to the puzzle's solution.

Most likely you experienced one of three responses to the dot problem. First, the solution was immediately obvious to you. (If so, my apologies.) Second, you may have thought about the puzzle and slowly, via step-by-step reasoning, come up with the correct answer: "I'm certain that the 'obvious answer' can't be the correct one. That means that the number of dots to be added must be less than eight, but by how many dots? Let's just close the figure and count the dots needed to do that." The third response may have involved an initial failure to come up with the solution, further pondering of the problem, and then a nearly instantaneous insightful resolution—the Aha! response.

Aha! is shorthand for the feeling that accompanies solving a problem with sudden insight and creativity. The prototype for this response is attributed to the Greek philosopher-scientist Archimedes. According to legend, the tyrannical king Hiero of Syracuse challenged Archimedes to find the solution to an unusual dilemma. Hiero suspected that an artisan had cheated him when fashioning one of his crowns by mixing silver with what should have been pure gold. Archimedes was charged with coming up with a test to allay or confirm the king's suspicions. But in doing so he could not do anything that would damage the crown. This requirement eliminated the usual chemical tests for distinguishing pure gold from gold mixed with an adulterant.

The answer occurred to Archimedes in his bathtub. While lolling in the water he noticed that not only did the water rise but his weight seemed to decrease. In an instant— the Eureka! or Aha! moment—Archimedes realized that two objects of equal weight will displace different volumes of water when they are immersed, unless their densities are equal. Because silver is less dense than gold, the gold-silver alloy imitation did not displace the same amount of water as a solid gold crown.

Today we can identify the processes that likely took place in Archimedes' brain as he achieved his insight (now known as Archimedes' principle). Studies using fMRI and EEG techniques identify a different response when problems are solved by sudden insight as opposed to step-by-step logical deduction. About three hundred milliseconds (thousandths of a second) before the Aha! moment, a sudden burst of activity appears in the right hemisphere maximal in the anterior temporal area. According to the interpretation of the scientist carrying out the fMRI study, this right-hemisphere activity stimulates indirect associations: "making connections across distantly related information that allows for seeing connections that previously eluded them."

Here is what happens in response to that right hemisphere stimulation: Ordinarily, our left hemisphere manages the everyday verbal-reasoning processes that we all use to solve problems not requiring creative solutions. But when this step-by-step approach fails to yield a solution—especially when a creative solution is called for—that high-activity burst from the right hemisphere projects to the left hemisphere, where it activates novel circuits that lead to forming the new associative linkages underlying the Aha! solution.

The importance of the right hemisphere for creativity was first discovered by means of a clever experiment. Problem solvers were given written hints suggesting the correct solution to a problem. The hints were placed on the screen so as to be read preferentially by either the left or the right hemisphere. Since the left hemisphere is specialized for reading and language in general, it would seem reasonable to assume that the hints would prove more effective when delivered there. But that isn't what the experimenters found. Hints are more likely to be helpful if they're read by the right hemisphere. At first this seems surprising, but upon reflection, it actually makes perfect sense: insightful solutions to problems require the creation of novel associations. The right hemisphere is specialized for this.

Here is an example of the right hemisphere providing a solution that eluded the efforts of the left hemisphere:

"What is the single word that can form a familiar compound word or phrase with *pine, crab,* and *sauce*?"

If you immediately come up with the answer (*apple*, which forms pineapple, crab apple, and applesauce), your left hemisphere will be doing most of the work. But if you have to work on it for several minutes before you suddenly "see" the answer via an Aha! experience, your right hemisphere will be preferentially activated, according to fMRI studies carried out by neuroscientists at Northwestern University.

In other words, solving a problem via the Aha! experience is qualitatively distinct from the usual left hemisphere–directed methods for problem solving. Creativity involves a shift in brain activation from left- to right-hemisphere processing. This allows the forging of novel associations. This shift from our usual left-hemisphere-dominant brain processing to the right hemisphere, with its novel patterns and circuits, is the basis for the pleasure we experience whenever we come up with a creative solution to a problem or puzzle.

So if you want to cultivate creative associations, it helps to shift the activation patterns of your brain toward right hemisphere processing. In addition, creativity is enhanced by the "power down" states of rest, relaxation, sleep, and especially dream sleep. All of these power-down states have a common mechanism: they all initiate changes in the ratios of the brain's neurotransmitter systems.

Other studies of the brains of creative people have emphasized the importance of the temporal lobes. R. E. Jung, a psychologist at the University of New Mexico, points to the association of creativity with specific degenerative diseases affecting the temporal and frontal lobes. Typically, an aged person who has never been particularly creative prior to the onset of fronto-temporal dementia may take up painting and

after little or no instruction produce highly original paint-
ings. So far no one has explained how brain disease, which
reduces brain performance in general, can lead to an increase
in creativity.

One theory claims that creative impulses exist in all of us
but are weakened by custom and social rules of behavior—a
variation of the claim dating back to French philosopher
Jean-Jacques Rousseau that formal schooling inhibits rather
than stimulates creativity. Consistent with this view, the
power of these restraining influences primarily from the frontal
lobes is reduced in frontotemporal dementia. The result is an
outpouring of creativity originating principally within the
temporal lobes.

In order to test this hypothesis, Jung recruited students
from his university and measured elements found to be
important to the creative process. One such element is the
ability to come up with a series of uncommon uses for a common object such as a brick (paperweight; displacer of water in
a toilet tank to save water with each flush). This test was
combined with free-figure drawings (the students were asked
to come up with as many drawings as they could in a limited
period of time) followed by a restrained-drawing exercise
(limiting the drawings to only a few lines), with both exercises aimed at creating original drawings. All of these tests
were then tallied to yield a composite creativity score.

Two areas showed greater activation on fMRI testing of
the brains of students who ranked higher in creativity: the
left temporal region (critical to the assessment of social concepts) and, to a lesser extent, the posterior cingulate (important in the identification of complex relationships).

As Jung admitted to me during a discussion, this work is only preliminary (as of late 2007). Over the next few years Jung plans additional research aimed at further refining our understanding of the relationship between creativity and the brain.

Despite the ongoing tentative nature of brain research on creativity, I believe a practical conclusion seems justified: creativity can be increased by divergent-thinking exercises.

The Power of Visual Thinking

In order to get a clearer idea about the relationship of creativity to divergent thinking (commonly referred to as thinking "outside the box"), I met with Temple Grandin at a brain conference in Tucson. The author of *Thinking in Pictures*, an account published in 1995 describing her life from the dual perspective of a scientist and an autistic person, Grandin now divides her time between serving as a professor in the Department of Animal Sciences at Colorado State University, where she specializes in livestock handling and behavior, and her second career as a much sought-after lecturer on Asperger's syndrome. Grandin is famous among cattle breeders for her creative designs for humane livestock equipment, and famous among a larger public for her many articles and several books about what life is like for someone afflicted with autism. Earlier that day, before our meeting, I attended a lecture on creativity that she delivered to an audience composed principally of psychiatrists and neurologists.

Dressed in a western shirt and boots, dark trousers, and a

string tie, Temple strode confidently across the stage while holding a single sheet of handwritten notes that she rarely referred to. She told the assemblage of neuropsychiatrists that people possess one of three "thinking patterns" or "specialist" brains.

"The first, and the most common in our culture, is the *verbal-language* brain. Typically, this includes writers, translators, and any other profession that doesn't place much emphasis on visualization. As a rule, people with this brain specialization do poorly in mathematics. The second, the *music and math* brain, is self-explanatory. These folks are your musicians, your computer geeks, your math geniuses." Temple includes herself in the third category: the brain specialized for *visual thinking*.

Words are only a "second language" to Temple, who converts whatever is said to her into "full-color movies, complete with sound, which run like a VCR tape in my head." Recalling a specific memory consists of replaying the experience "as if it were on a CD-ROM disc." She then uses the images from her visual memory to help her solve design problems for cattle chutes, truck loading, and sorting facilities.

As an example of Grandin's point about the primacy of images in the creative process, read the following paragraph, and after recalling as many details as you can, write a short narrative based on it:

"If the balloons popped, the sound wouldn't be able to carry, since everything would be too far away from the correct floor. A closed window would also prevent the sound from carrying, since most buildings tend to be well insulated. Since the whole operation depends on a steady flow of electricity, a break in the middle of the wire would also cause problems. Of course, the fellow could shout, but the human voice is not loud enough to carry that far. An additional problem is that a string could break on the instrument. Then there could be no accompaniment to the message. It is clear that the best situation would involve less distance. Then there would be fewer potential problems. With face-to-face contact, the least number of things could go wrong."

Most people are able to remember only about 20 percent of the information contained in that paragraph. That's because even though each sentence makes perfect sense, the overall meaning of the paragraph remains obscure. What is being described? Without an answer to that question, creativity cannot be harnessed to produce an original narrative. Yet if you look for just an instant at the cartoon on page 185, the paragraph makes perfect sense and you will have little trouble fashioning a short narrative, in this case a scene from a classic love story.

Artists, inventors, and scientific innovators speak often of an image that inspired them. Einstein arrived at his theory of relativity by imagining how the movement of a person in one train would appear to an observer located in a second train running parallel to the first. On pages 175–76 we described Kekulé's discovery of the chemical benzene, which evolved from his memory of a gold ring fashioned in the form of the alchemical symbol for the unity and variability of matter— two intertwined snakes biting their own tails. In each instance, a vivid image provided the material for creativity by temporarily counteracting the inhibiting effect language exerts on our capacity for visual thinking.

In modular design, whether in the writing of a novel or in the direction of a movie, the creator uses flashbacks and other narrative devices rather than telling the story from beginning to end in a strictly linear manner. "Modular design is an attractive way to show relationships between events or people or motifs or themes which are not generated by sequences of cause and effect and so are somehow atemporal, perhaps even timeless," says novelist Madison Smartt Bell.

How to shift one's mental balance from a chronological to a modular design? Novelists such as William Gibson (*Spook Country; Pattern Recognition*) and Michael Ondaatje (*The English Patient; Anil's Ghost*) and filmmakers including Christopher Nolan (*Insomnia; The Prestige; Memento*) do it all the time. Instead of working from a narrative progressing from past to present to future, these artists imitate the method of the mosaicist who assembles fragments of glass and tile to form what can be seen, at a greater distance, as a coherent form. This *modular design* method frees the writer or filmmaker from the limitations of *linear design*, which concentrates on the overall movement of the narrative from its start to its finish.

All of us use modular design on a regular basis. For instance, if I ask for your opinion about someone you have known for a long time, your answer won't be based on reviewing in your mind an imaginary video arranged according to a linear chronology of your experiences with that person over the years. Instead, you'll base your answer on fragmentary, short vivid images—jump shots, in cinematographic terms—that link your impressions, observations, and experiences with that person. Your final "sense," your conclusion about that person, will depend on how you arrange and rearrange these fragmentary observations and equally fragmentary facts. In order to do this, you fashion a "modular narrative design," where, as Madison Smartt Bell describes it, "narrative elements are balanced in symmetry as shapes are balanced in a symmetrical geometric figure, or as weights are balanced on a scale."

One way that I use modular design as a creativity tool is to quickly and uncritically write down on a large sheet of drawing paper everything that occurs to me about a specific person or event. I then shift these separate observations around in random order and see what structure emerges. If I'm near a computer, I prefer to use the innovative program Inspiration: The Visual Thinking Tool, which helps me to perceive hidden relationships that evade casual inspection and often spur creative ideas.

Four Steps to Increase Creativity

Recent research suggests four practical steps that can be taken to increase creativity.

1. Focus on the problem for as long as you need to understand it.
2. Mentally put into words your implicit assumptions. Write a summary of your understanding of the problem.
3. Make certain that you understand what you must do to reach a resolution.
4. Ask yourself, "In what other ways can I envision this problem?"

When the answer still isn't forthcoming, put it out of your mind, and cultivate one of the power-down states of rest, relaxation, sleep, and, especially, dream sleep.

If you take these steps you will be applying the four main

characteristics of divergent thinking first identified in the 1950s by psychologist Joy Paul Guilford: *fluency*, rapidly producing multiple possible solutions to a problem; *elaboration*, thinking through the details of an idea; *flexibility*, entertaining multiple approaches to the problem simultaneously; and *originality*, coming up with ideas that don't occur to most other people.

Here is a puzzle that exemplifies creative thinking. I learned about it from puzzle developer Dave Youngs; it can be solved only by engaging in divergent thinking.

Six drinking glasses are arranged in a row. The first three are filled with water; the next three are empty. How would you get the full and empty glasses to alternate by moving only one glass? You can work at this by imagining the arrangement of the glasses in your mind or by actually setting out the glasses. Trust me, it's possible to alternate the full and empty glasses by moving only one of them, but in order to do this you have to think divergently.

This is the reasoning process used by one person who solved the puzzle: "Perhaps I could eliminate one glass altogether [he tries that]. . . . That doesn't do it. . . . Maybe I could move one glass and then moving it back to its original position . . . but that wouldn't work, either, unless the one glass could be moved. . . . Aha!"

The correct solution to this puzzle requires the solver to overcome two limiting assumptions. First, that the word *moving* necessarily implies rearranging the glasses. Second, that moving one of the glasses must be carried out only in the horizontal plane. Buying into either of these assumptions prevents a successful solution to the puzzle: the alternating lineup of the glasses cannot be achieved by simply sliding on the flat surface of the table one of the glasses to another position. But by *vertically* lifting glass two and empting its contents into glass five, and then moving it back to its original position the alternating arrangement of full and empty glasses is achieved.

Notice that the correct solution involves all four of the steps I suggested, along with Joy Paul Guilford's four characteristics of divergent thinking. Most important was step four: Ask yourself, "In what other ways can I envision this problem?" (Corresponding to Guilford's *originality:* coming up with ideas that don't occur to most people.)

Here is another puzzle that I doubt you will solve easily unless you have encountered it before. It cannot be solved without a decisive revision of one very specific limiting assumption:

One costs fifty cents, you can buy twelve for one dollar, and 312 for one dollar and fifty cents. What objects are being bought?

Keep in mind that the solution calls for *objects*, so you can dispense with any solution involving units of time (minute allowances of cell phone calling plans, etc.). Since this is a much more difficult puzzle than the water-glasses puzzle, here's a hint in the form of a conversation between a customer, Ned, and Susie, the owner of a specialty store.

Ned: "I'd interested in buying some of these. How much does one cost?"

Susie: "One would be fifty cents."

Ned: "Suppose I buy twelve?"

Susie: "Twelve would be one dollar."

Ned: "Okay, I would like 312."

Susie: "That will bring your total to one dollar and fifty cents."

This puzzle's difficulty stems from the fact that such capricious pricing doesn't make sense according to the rules of commercial exchange. If one item is worth fifty cents and twelve items cost only fifty cents more, the buyer has gained eleven items that he can sell for fifty cents each. The situation becomes even more preposterous with the purchase of 312 items for only $1.50. Such a scenario isn't possible in regard to items in the usual meaning of the term *item*. But what kind of item might be sold in units that would make one cost fifty cents, twelve cost a dollar, and 312 cost $1.50. Get it now? If not, here is a final hint:

Ned has just moved into a new housing development and is buying items that will enable the guests coming to his housewarming party to find his not-quite-finished house. Got it now? Ned is purchasing the address numbers 1, 2, and 3 from a hardware store so that he can place them to identify his new house at 312 Westover Street.

In both of the above puzzles the answer required "thinking outside the box"—an informal term for divergent thinking. Indeed, you can solve the following Nine Dot Puzzle only by thinking outside the box, *literally*.

On a piece of paper draw nine dots as shown on page 191. Now connect each of the dots with four straight lines. Each dot must be crossed once, and only once. You may not lift your pencil from the paper once you start.

After a few unsuccessful tries you'll likely conclude that

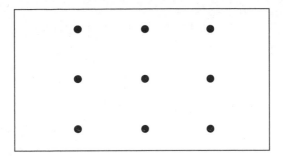

no fewer than five lines are needed to include all of the dots. But read the instructions again while keeping in mind the second and fourth step in the formula to increase creativity: "Mentally put into words your implicit assumptions" and "In what other ways can I envision this problem?"

The only explicit requirements were: (1) that you not lift the pen or pencil from the paper once you begin, and (2) that each dot not be crossed more than once. Notice that the instructions did not state that you couldn't extend the lines beyond the arrangement of the dots to solve the puzzle. If you envision the problem without that unwarranted assumption, one of the many solutions will occur to you. (Page 194 shows one solution of this puzzle, which, according to legend, served as the inspiration for the term "thinking outside the box.")

Here is another puzzle, the Shrinking Square Challenge, created by Dave Youngs. Place four pennies on the corners of the square drawn on page 192. By moving only two of the pennies, create a square that is smaller than the original square. One solution to the puzzle is shown on page 195.

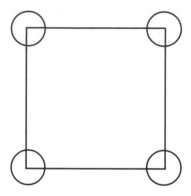

As a final example of creative, outside-the-box thinking, look at the diagram on page 193. It depicts the famous outdoor maze at Hampton Court in England. Although the maze doesn't present much of a challenge on paper—one can simply trace one's way through it with a pencil—tourists regularly become disoriented once inside the maze. For one thing, after they make only a few turns, it becomes difficult for them to determine where they are and what direction they are facing. Second, after entering the maze they are prevented from seeing farther than the next passage by high, bulky hedges. The result is an escalating panic and a sense of disorientation. The bewildered tourist would be surprised to learn that he would have done better if he had closed his eyes when entering the maze.

The Hampton Court maze can be traversed by putting out your right hand and touching the hedge on the right as you enter. If you maintain that contact as you proceed through the maze, you are essentially walking along the walls of a very large, irregularly shaped room. Eventually you will reach the center of the maze and then return to your starting point.

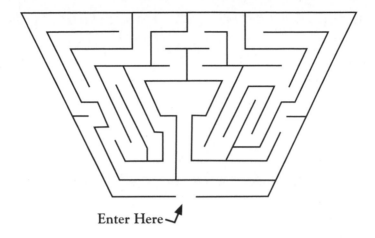

Enter Here

Notice that sight plays no part in successfully negotiating the maze; in fact, the more you rely on vision, the greater your chances of getting lost.

One final point. Some of us aren't innately drawn to solving problems and puzzles. While it's true that they sometimes demand a lot of time that could be applied to work or other more "important" pursuits, they can enhance brain function in ways that work or leisure pursuits rarely do. They provide a fun means of strengthening concentration by cultivating the habit of mulling over a problem, thinking about it at odd times of the day, and pondering it that evening in the waning moments of consciousness just prior to sleep. Most important, they integrate brain function, uniting the right and left hemispheres. It's both pleasurable and brain-enhancing when the solutions require a fundamental shift in one's mental perspective—resulting from activation of our right hemisphere. If you work at puzzles such as the Nine Dot Puzzle and the Shrinking Square Challenge, you will strengthen your capacity to question initial assumptions, restate and reframe problems, and increase right and left temporal hemisphere activity. The result will be a greater ability to come up with creative solutions.

SOLUTIONS

Answer:
The next figure contains 32 dots.
At each step a new side is added to
the square. Completing the last side
adds only 4 dots, not 8.

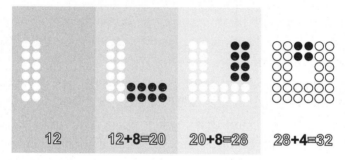

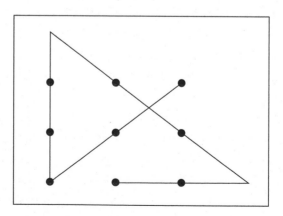

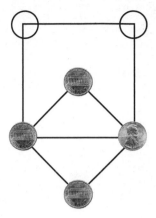

PART SIX

Impediments to Optimal Brain Function, and How to Compensate for Them

D espite its impressive processing power, the brain operates under severe constraints whenever we attempt to do more than one task at a time. Of course it's *possible* to simultaneously drive and talk on the phone while monitoring a backseat conversation, but such multitasking comes at a price: increased response time and decreased overall efficiency—and on the highway that translates into more accidents.

Like it or not, our brain was designed for efficiently processing one cognitive operation at a time. This is especially true when decisions are involved.

According to researchers at Vanderbilt University, listening and responding to sentences while mentally rotating pairs of geometrical figures reduces overall brain activity by almost 30 percent, as measured by fMRI. This decrease is linked to an overall decrease in efficiency: it takes longer to do each task. Loss of efficiency occurs especially when a second task is begun before completing the first.

The explanation for such delays was discovered by the Vanderbilt researchers using fMRI. A "neural network" within the frontal lobes acts as a "central bottleneck of information processing that severely limits our ability to multitask," according to the Vanderbilt neuroscientists.

If decisions aren't required, that bottleneck isn't created. For example, many surgeons claim that operations go more smoothly if they listen to music in the operating room. Since no decisions are required about the music (except for choosing the selections before the operation), the surgeon can concentrate all of his brain resources on surgery and that first negative consequence, the bottleneck, is avoided.

To appreciate the second negative consequence of multitasking, consider e-mail and instant messages, currently the two biggest contributors to our multitasking culture. A study involving Microsoft employees found that it takes about fifteen minutes for a worker to return to mentally challenging work (writing computer code, for example) after answering incoming e-mail or instant messages. Many of the workers became so distracted that they spent additional time checking favorite news, sports, and entertainment websites.

Another study on worker productivity found that 28 percent of an office worker's day was spent dealing with interruptions and the resultant recovery time needed to redirect attention to the main tasks. According to RescueTime, a company that analyzes computer use in the workplace, the typical computer-based information worker checks e-mail fifty times a day and sends or receives seventy-seven instant messages. Many of these are unnecessary, distracting, and disruptive of sustained attention. The resulting losses in productivity are estimated to be as high as $650 billion a year. So urgent has this matter become that Microsoft, Intel, Google, and IBM, among others, formed in the summer of 2008 the nonprofit Information Overload Research Group, aimed at coming up with solutions.

Third, and most important, multitasking creates stress.

Think about the last time you tried to monitor a basketball game on television while talking to a friend on the telephone. Such a situation was stressful because you were recruiting the same brain areas to perform two competing tasks. Your brain was in conflict with itself: you were trying to watch and listen to the game while also trying to listen to your friend's remarks. To reduce the stress of the situation, you had several alternatives. You could have recorded the basketball game to watch later or told your friend that you would call her back. As a less efficient alternative, you could have turned off the sound and simply watched the action while continuing your conversation. If you had done that, you would have been slightly less distracted because the language centers for talking on the phone and the visual processing areas for watching the game call upon different brain resources.

As we age, we get worse at multitasking. In one of Arthur Kramer's studies on multitasking and age, he monitored the performances of both young and old drivers as they took part in a simulated driving test while carrying on a conversation. He found that while all of the participants tended to miss unimportant variations on the simulator, older drivers often failed to notice and remain attentive to important variations of the simulated scene that could lead to an accident.

In order to avoid stress-associated multitasking, whatever your age, learn to rivet your attention on what you are doing at the moment. This holds true both at work and at leisure. "Practice only as long as you can concentrate," teaches famous PGA golf coach Jim McLean, author of *The Eight-Step*

Swing, one of the most popular golf instruction books ever published. "Stop if you lose focus. Short, focused practice sessions are the most productive." Golf legend Ben Hogan advocated a similar approach. "Adopt the habit of concentrating to the exclusion of everything else while you are at the practice tee and you will find that you are automatically following the same routine while playing a round in competition."

No doubt you've noticed that multitasking shares features with working memory. But multitasking differs from exercising your working memory in several important ways:

Multitasking usually involves time urgency: stressful attempts to "save time" by doing two or more things simultaneously. As mentioned, this is doomed to failure; it's better to complete one activity before starting another. Working memory isn't about saving time but improving performance by holding online several independent lines of thought—as with a musician practicing the art of counterpoint. Novelists routinely exercise their working memory by skillfully blending dialogue, narrative, character, mood, and writing style in order to bring everything into balance. While describing the action of a character, for instance, the novelist must retain in working memory all of the things he has written earlier about her.

From the brain point of view, multitasking is inefficient at best. When you're multitasking, you're going to make more mistakes, mistakes that can be costly and potentially dangerous. At the very least, you're going to feel increasingly stressed and harassed if you establish the multitasking habit. As an antidote, slow down, take a few deep breaths, select which of competing tasks is most important and do that one first (prioritize), and then finish that task before moving on to something else.

Compensating for Age-Associated Brain Changes

When he was eighty years old the concert pianist Arthur Rubinstein was asked in a television interview how he managed to retain his level of expert playing. He responded that he played fewer selections, practiced them more frequently, and played the slower segments even more slowly than usual in order to make the ensuing fast segments seem even faster.

The novelist Harriet Doerr described a similar acceptance and compensation for age-imposed limitations in an interview I conducted with her five years before her death at age ninety-two. Doerr, who didn't start serious writing until she was sixty-five, won the National Book Award in 1984 at age seventy-three for her novel *Stones for Ibarra*. When I asked about the effect on her writing career caused by a severe late-life-onset visual handicap (macular degeneration), she responded:

"Adjusting to my loss of vision has been hard, but I have enjoyed some secondary benefits. When you can't see, you think a lot more. The best exercise of my imagination is sometimes lying in bed a few moments in the morning in the silent house and letting things like a difficult sentence or the ending of a story simply float into the mind. I've also learned to employ various methods of substitution for my visual handicap, such as my recent switch to a word processor."

Rubinstein and Doerr are good examples of the three-part strategy for overcoming some of the limitations imposed by age that were first suggested by the late Paul Baltes, the main investigator for the Berlin Aging Study: selective optimization with compensation (SOC).

Selection: Acknowledging rather than denying an area of weakness (Rubenstein's decision to concentrate on fewer pieces; Doerr's acceptance of her visual limitations).

Optimization: Working harder to overcome limitations (Rubinstein's increase in his hours of practice; Doerr's early-morning efforts to think "a lot more" about her stories).

Compensation: Using new ways to counteract decline (Rubinstein's employment of a kind of impression management whereby fast passages seem faster than they actually are because they are preceded by exaggeratedly slow passages; Doerr's shift from writing by hand to using a computer).

While Bates's SOC strategy theory can be applied by anyone over forty years of age, it's especially helpful for those over fifty-five. If you're fifty-five or older, the odds are that you are experiencing limitations in your speed of information processing and your working memory. What's more, these functions will continue to decline as you move into your sixties and beyond.

The good news is that the deterioration isn't the same for everyone—an indicator that the brain ages at different rates and that a general decline isn't an inevitable result of aging. That's why it makes sense to remain positive and learn to work with the age-associated changes that accompany longevity.

One thing we do know for certain: A healthy brain is a prerequisite for success, even survival, during our mature years. Therefore it's important to understand how it differs from the brains of young to middle-age adults.

The Strengths and Advantages of the Mature Brain

During my brain lectures I'm often asked if our IQ lessens as we get older. The answer depends on which of the two kinds of intelligence one is referring to. The first, *fluid intelligence*, can be thought of in computer terms as the hardware of the brain that determines the sheer speed and accuracy of information processing; it includes any activity involving the registration and use of new information. When we try to learn a new language or accommodate to new demands at work, we're calling upon our fluid intelligence. Reaction time, attention, concentration, and working memory are all components of fluid intelligence—and they all decrease in efficiency as we age. This age-associated falloff in fluid intelligence is believed to result from a loss of neurons in the subcortical nuclei—the juice machines—especially the nucleus basalis, which manufactures and distributes the neurotransmitter acetylcholine. It's the decrease in acetylcholine and other neurotransmitters that leads to a decrease in enthusiasm and general "get up and go" energy. It also leads to distraction and difficulty "sticking to" a project or line of thought.

Crystallized intelligence (CI), in contrast, doesn't lessen with aging. Sometimes referred to as the "software" of the brain, it can be compared to the computer programs that we install on our PC over the years. These vary according to cultural rather than biological factors: where we grew up, our reading and writing abilities, our language, the schools we attended, our educational attainments, and professional skills. CI also includes one's mastery of emotional and social skills. All of these components of CI will vary from person to person, depending on life experiences. But as with Arthur Rubinstein's piano virtuosity and Harriet Doerr's writing genius, performance based on CI will remain largely unaffected in most people as they age (assuming they are free of brain disease).

In short, maturity induces a kind of standoff within the brain. The impairment of fluid intelligence makes it difficult for an older person to compete with someone younger on tests involving the rapid retrieval of information—one of the reasons people older than fifty-five rarely appear on *Jeopardy!* But when rapidity of response isn't required, the older brain is often even more proficient in some areas than its younger counterpart.

Think for a moment about what you're usually looking for when you seek help with a complicated professional or personal problem. It usually isn't quick analysis (the product of fluid intelligence); truly challenging problems usually aren't quickly solved. Instead, you're looking for someone possessing what Paul Baltes defines as wisdom: "a state of knowledge about the human condition, about how it comes about, which factors shape it, how one deals with difficult problems, and how one organizes one's life in such a manner that when we are old, we judge it to be meaningful."

Because they have lived long lives, older people have had the opportunity to see and learn more. As a result, they are in a position to synthesize knowledge in many different areas and provide it to others in the form of wisdom. At the moment, scientists lack an objective measure of wisdom. Nor is psychological testing very helpful here. A twenty-five-year-old may achieve impressive scores on tests of memory, fluid intelligence, and general mental performance; yet compared with an older, more experienced person, the twenty-five-year-old would probably make a less capable judge, surgeon, or airline pilot—three professions requiring years of experience before a person reaches the highest level of achievement.

"Experience," "maturity," and "judgment" are among the terms frequently employed in industry, science, and the arts to describe the wisdom-based performance of the mature worker. Another synonym is "expertise," consistently superior performance in a specialized area. Examples include retained levels of expertise over the life span of professional musicians, airline pilots, surgeons, and trial attorneys. In most instances, this expertise is restricted to their area of proficiency. An older concert pianist, for instance, shows the same age-related slowdown in brain processing as other people of the same age do. Only musical ability remains largely unaffected. (Notice that I've slyly introduced the qualifier "largely" in the previous sentence because, as with the methods described by Arthur Rubinstein, the aging pianist often compensates for age-related declines in brain function by resorting to various "tricks" that are noticeable only to other musicians and critics. Indeed, the key to retained performance and successful brain aging is to learn and practice such "tricks.")

As an example, consider transcription typists. Despite a general slowdown in reaction time and speed of hand responsiveness, a skilled older typist's performance may be comparable to that of a typist twenty or thirty years younger. How do they do it?

In a classic 1984 study, the renowned psychologist Timothy Salthouse tested the reaction time and typing skills of typists of all ages. He discovered, as anticipated, that the general reaction times of older typists were generally slower than those of their younger counterparts. Yet they typed just as fast. Puzzled at this unexpected discrepancy, Salthouse considered two possibilities. Perhaps the older typists had been even faster at an earlier period in life and had simply slowed down. Or maybe the older typists were using a different strategy.

In order to decide which explanation was more plausible, Salthouse retested his subjects, but this time he imposed a crucial limitation based on an elementary principle about typing. The superior speed of expert typists is related to how far they can look ahead in the text beyond the word they are currently typing. As their skill level increases, expert typists prepare for upcoming keystrokes by scanning ahead in the text and anticipating future key strokes with a premonitory moving of the relevant fingers toward the required location on the keyboard.

When Salthouse restricted the typists from looking ahead, the younger typists' speed remained about the same, while the older typists' speed decreased dramatically. A similar look-ahead strategy holds for older pianists when they sight-read pieces of music.

As a result of changes in technique, older, more experienced architects, graphic designers, and others whose work involves the exercise of spatial ability perform comparably to younger workers. Older expert pilots perform as well as younger pilots on decision-making challenges because their extensive experience and knowledge enable them to quickly home in on the most critical information needed to solve the problem. Another compensatory strategy for older pilots is to fly fewer but longer routes, thus reducing the number of takeoffs and landings, which contribute to both stress and workload.

Overall, expertise neutralizes age-associated performance differences. This fact leads to an eminently practical application: as you get older, you can better your odds of remaining active in the workplace by enhancing crystallized intelligence.

As you reach your sixties and seventies, your success will become increasingly dependent on what you know and how well you can apply that knowledge, especially if you possess special skills or expertise that distinguishes you from other workers.

Maintaining Brain Fitness at Work

Recently leaders of major corporations have become interested in finding ways to increase the brain fitness of their employees. This interest stems in no small part from the challenge of a graying workforce.

Ours is the longest-lived generation in history. Until about two hundred years ago the average human life span was about thirty years. During the twentieth century that average life span doubled. Nor is there any sign of its slowing down. It has been increasing at the bewildering rate of 2.2 years per decade (or five hours a day) for the last hundred years. The impact of this life span expansion on the workforce is certain to be profound. By 2010 the number of U.S. workers ages forty-five to fifty-four will grow by 21 percent; the number of fifty-five- to sixty-four-year-olds, 52 percent; and the number of workers sixty-five and older, 30 percent.

Nor can employers rely on a dependable source of younger workers to take the place of aging ones. Thanks to differences in what's been termed "generational knowledge transfer," younger workers, as a rule, are less likely than their older counterparts to remain permanently at the same company.

Thus corporations are hit with a double whammy: older workers are retiring, while the younger ones can't be persuaded to stay. In response to these trends, businesses are now interested in increasing the number of employees who keep working beyond the usual retirement age. But in order to do that they have to learn how the brain of the older worker differs from his counterpart who is twenty or thirty years younger.

The table on page 211 depicts the four generations currently working alongside one another. Each has operational styles fashioned at least in part by critical events that occurred in their lifetimes (World War II, the Cold War, the

KEY WORKPLACE TRAITS
of each generation

Mature/Silent	*Baby Boomers*
• Long tenure with organizations • Respect hierarchies and authority figures • Like structure and rules • Demonstrate strong work ethic • Pay attention to quality of work • Less mobile	• Skeptically "accept" authority figures • Are results-driven and ambitious • Have long-term aspirations with organizations • Retain what they learn • Are idealistic and competitive • Are people-focused • Optimistic overall
Generation X	*Generation Y*
• Comfortable with diversity • Value freedom and informality • Have short-term loyalty • Work well in networks and teams • Embrace technology • Seek work/life balance • Learn quickly • Skeptical overall	• Comfortable with diversity • Value informality • Have short-term loyalty • Learn quickly • Embrace technology • Need supervision

civil rights movement, the women's movement, gay rights, and advances in communication technology). Notice that two traits of the Mature/Silent workers are likely to bring them into conflict with members of the other three generations: their respect for hierarchies and authority figures, and their preference for structure and rules.

If you're sixty years of age or older, you won't be able to count on doing things in the workplace in "traditional" ways. Instead, you will be working with coworkers who, on the whole, bring very different styles, attitudes, and skills according to their age. So if you plan to keep working beyond age sixty, you will have to adapt to the patchwork of operational styles that characterize the current workplace.

Fortunately, many corporations now recognize the value of retaining mature workers. In order to avoid the loss of what is called "institutional memory"—shorthand for how the company gets things done—many of them are adopting new policies aimed at retaining valued older employees. One retailer's recent recruiting poster reads: "Now hiring WISDOM."

Institutional memory and wisdom are components of crystallized intelligence—the intelligence that isn't affected by aging. This includes brain networks involved in expert knowledge, among others. If you're older and want to remain employable in the twenty-first-century marketplace, establish an expertise that distinguishes you from your younger coworkers.

Despite your own wishes on the matter, you will eventually reach an age when retirement will be forced upon you (unless you're self-employed). At the moment, almost all organizations enforce mandatory retirement by seventy at the latest. The challenge then involves transferring and incorporating the brain-enhancing activities you've developed over a lifetime to your retirement years.

Retirement: A Concept in Need of Redefinition

From the point of view of brain health, traditional retirement doesn't make sense. Instead of recreation, anyone contemplating retirement should be seeking *re-creation:* finding new and healthy ways of challenging the brain by continuing to apply one's natural talents and skills. As an example of what can happen when that isn't done, consider the experience of Roger, a recently retired lawyer I know (his name has been changed to preserve anonymity).

During a successful legal career Roger spent hundreds of hours in meticulous research and artful analysis of his cases, both within and outside the courtroom. This persistence had served him well. at retirement he was a senior partner and highly regarded by the associates in his firm for his thoroughness.

Finding himself with nothing to do during the first few weeks of his retirement and with unlimited time on his hands, Roger decided for no pressing reason to commission a re-survey of the property line separating his land from his neighbor's. This seemed curious even to his own family, since Roger had never expressed any curiosity about the property line during the forty years the family lived in the house. When the surveyor reported "perhaps" a one- to two-foot-wide swath of land that wasn't clearly assignable to either property, Roger began a fractious disagreement with his neighbors over the need to establish the "exact" boundary between the two properties.

While both the neighbors and Roger's family agreed that something should probably be done to clarify the boundary, why not simply mutually decide on a new boundary line at the halfway point and be done with it? But such a casual approach didn't soften Roger's resolve, despite the likelihood that his intransigence would force a confrontation with neighbors with whom he had maintained friendly relations for decades. Everyone was puzzled and searched for an explanation for Roger's sudden interest in firmly establishing property boundaries.

The most likely explanation, I believe, is that Roger's careful, methodical, and meticulous approach to his work—an asset in his profession—turned into a liability after his retirement. Finding himself with too much unstructured time to fill, and deprived of an outlet for his preoccupation with accuracy, Roger applied his obsessive and compulsive traits to a question that demanded flexibility and tact. His rigid approach to the boundary question alienated and angered his neighbors, who, in turn, responded with more emotion than usual for them, largely because of Roger's petty exactitude and lack of accommodation.

Like Roger, most of us will be bringing to retirement the same personality traits that served us well or ill during our careers. That's why it's important to find some postretirement activity that lets us express these traits in healthy ways. For instance, a retired surgeon I know has shifted his interest since retirement from surgery to his hobby: the management of his stamp and art collections. He can now apply the exactitude he had during his surgical career to his collections.

The failure to find positive ways to use old habits and propensities accounts, I believe, for much of the dissatisfaction and unhappiness of retirees. Changing these habits is possible, of course, but probably unrealistic for most people: habits formed over five or six decades aren't likely to disappear the month after retirement. Rather than attempt to eliminate socially troublesome traits such as competitiveness and compulsiveness, retirees should seek out new ways of expressing them in a socially acceptable manner. The passion for golf, tennis, bridge, and other "competitive" activities at clubs or in retirement communities serves this purpose, I'm convinced. They offer the opportunity to channel lifelong competitiveness in ways that won't lead to social friction. So if you're thinking of retiring, first take stock of what you liked about your work and the personality traits that were important to your success. Then think of ways to apply these traits during your postretirement life.

Developing a Balanced View of Alzheimer's Disease

To manage the complexities of twenty-first-century life we need to retain our mental powers throughout the course of our lifetime. That's why Alzheimer's disease strikes a particular note of fear in industrialized countries such as ours, where great emphasis is placed on the clarity and vigor of one's thinking.

Alzheimer's disease embodies two of our greatest fears: a loss of control and a dependency on others. Think of all the television commercials for investment firms you've seen that feature smiling past-middle-age couples strolling hand in hand in front of their beachfront retirement homes. "Security, independence, and total control of your life after retirement should be your goal—and you'll reach it by investing now with . . ." is the stated or implied message.

But since the brain is susceptible to cognitively impairing diseases at all stages of life, optimal brain function can't be *guaranteed*, no matter what measures you may adopt. "If brain function becomes impaired, it's the result of disease, not age," says John Morris, lead investigator of the Memory and Aging Project, Washington University School of Medicine in St. Louis.

Morris's point is something that I emphasize to my patients every day. Although we can do much to lessen our chances of coming down with Alzheimer's disease or other dread dementias, there are no guarantees. Although lifestyle changes can be effective in maintaining brain health, genetics still plays a role. Some people are more at risk for dementia, probably secondary to genetic defects that neuroscientists are homing in on. So, given all this, what is the best attitude and strategy to employ under these circumstances?

Unless you come from a family with a high rate of Alzheimer's or another late-onset brain disease, it's not unreasonable to think positively. The prevalence of

Alzheimer's is only about 1 percent at age sixty-five. It doubles every five years after that, to between 16 and 25 percent of Americans at eighty-five. In the United States about 46 percent of people over eighty-five are thought to have at least early Alzheimer's disease. While these are not totally reassuring numbers, why not look at the situation from the positive point of view? In fact, the number of people free of dementia, even among the oldest old (ninety or higher), exceeds the number of those afflicted with it. So even if you live well into your nineties, you are still more likely than not to be Alzheimer's free.

In addition, milder, slower-progressing forms of Alzheimer's are the rule rather than the exception. Many people with Alzheimer's continue to work at less demanding jobs for some time after diagnosis. Further, currently available drugs provide some measure of improvement if not cure in a significant number of people affected with mild forms of the disease. Several very promising drugs and other treatments such as vaccination are in clinical trial.

Most important, we can reduce the likelihood that these pernicious alterations occur in our brain by making the lifestyle changes described throughout this book. This isn't intended to deny the seriousness of Alzheimer's disease, but to provide some perspective—and some hope.

Building Up Cognitive Reserve

Twenty years ago, pathologists came upon an unexpected finding when examining the brains of people who had died without any signs of Alzheimer's disease. In order to highlight their findings, here is a one-paragraph review of the brain changes typically found in Alzheimer's:

If you look through a microscope at brain tissue taken from a patient who died with Alzheimer's disease, two findings stand out. First, up to a quarter of the neurons in the cortex (the 3-millimeter rim of tissue covering the cerebral hemispheres) are reduced to dense, intricately tangled bundles that look like the rusted cables of a long-sunken ship. Second, you'll see clusters of degenerating nerve endings enclosing a homogenous central core that when viewed with the aid of special chemical stains resemble ink smudges. The German neurologist Alois Alzheimer, after observing these tangles and plaques, first pointed out that they could serve as a shorthand marker for the mind-destroying brain disease named after him.

Now for that unexpected finding. At autopsy the researchers discovered Alzheimer's changes in the brains of a coterie of deceased elderly patients who had functioned perfectly normally even during the last years of their lives. Even though they had functioned without any signs of Alzheimer's disease, their brains contained the high plaque and tangle counts usually associated with that disease.

Investigation of these exceptions to the plaque-tangle-dementia association turned up a common trait: increased levels of education. In general, people with more education were less likely to have been diagnosed with dementia during their lifetime. Nor did the subject matter of their education seem to matter. A degree in physics provided as much protection as one in medieval history.

In an attempt to explain such unexpected findings the pathologists speculated, I believe correctly, that education in general makes for a more efficient use of brain networks, resulting in a greater ability to withstand the so far incompletely identified insults to the brain responsible for Alzheimer's dementia. The pathologists even came up with a term for it: *cognitive reserve* (sometimes called cerebral reserve).

As a useful analogy, think of cognitive reserve in monetary terms. The more money you have, the easier it is for you to manage financial downturns; compared with your less affluent neighbor, you will have to lose more money before you're financially wiped out. That analogy also helps explain a companion observation about education and Alzheimer's disease (or any other dementia, for that matter). When the person with greater education develops dementia, the disease has reached a more advanced state than would ordinarily bring on symptoms in a person with less education. What's more, death occurs soon after symptoms first appear. It's speculated that in such instances the afflicted person's cognitive reserve, initially higher than the average person's, finally ran out.

According to the work of Yaakov Stern of the Cognitive Neuroscience Division of the Taub Institute at Columbia University in New York City, greater cognitive reserve is linked with enhanced activation of a brain network in the frontal lobes. In both younger and older subjects, the higher their IQ scores, the better they did on memory tests and the greater the activation of their frontal lobe network. And since higher IQ scores generally correlate with more years of education, Stern's findings provide yet another reason for building up cognitive reserve by staying in school longer.

While the emphasis on formal education as an explanation for cognitive reserve seems to make intuitive sense, I think it's overly simplistic. If you think carefully about the cerebral reserve hypothesis, why would greater cognitive reserve necessarily be linked with *formal* education? It's just as likely that those same brain networks that can be brought to maximum efficiency by formal education can also be enhanced by informal education—what was once referred to as the school of hard knocks, which confers advanced degrees in street smarts.

For instance, an eighty-two-year-old patient of mine who quit school after the eighth grade earned millions of dollars over a lifetime of shrewd real estate investing. Or consider the late humorist Art Buchwald, who ranks close to the top of my personal list of the smartest people I've known. His formal education? Two years of college. Such exceptions to the education–cognitive reserve rule are rarely cited when learned professors publish papers lauding the brain benefits of higher education. No mystery here: the professors carrying out the studies possess doctoral and other advanced degrees and consequently are only too ready to equate intellectual abilities and a highly functioning brain with formal education. After years of associating with others with similar educations they are more than slightly uncomfortable with the thought that perhaps the mechanic who services their Lexus may be smarter than they are.

"Cognitive reserve is not something you were born with," Yaakov Stern observes. "It's something that changes and can be increased over time. Even something as simple as keeping up a regular reading schedule can increase cognitive reserve and reduce your chances of getting Alzheimer's disease."

Increase your cognitive reserve by engaging regularly in some form of cognitive training. According to Michael Marsiske, a principal investigator of a study funded by the National Institutes of Health that examined the effect of cognitive training on everyday functioning, mental decline among middle-age and older adults can be prevented by as little as ten sessions of mental exercises aimed at boosting reasoning skills, memory, and rapid mental processing.

When discussing cognitive training, Marsiske, a youngish-appearing forty-one-year-old neuroscientist, speaks volubly and with an infectious enthusiasm on his favorite topic: how to improve mental performance in middle to late adulthood through mental exercises. He described for me the mental exercise program he has developed as a result of his research:

For *memory training* he suggests forming visual images and mental associations as a means of recalling words and verbal narratives such as stories. This is similar to the methods described in Part Three. In order to develop *reasoning training,* he proposes such exercises as learning to discern the pattern in a letter or word series (e.g., *a c e g i . . .*) and then identifying the next item in the series (*k*—every other letter is omitted). *Speed of processing training* involves visual searching combined with divided attention: identifying an object on a computer screen at increasingly brief exposure times while dividing one's attention between two separate search tasks. For instance, judging which configuration of objects (two cars, two trucks, a car and a truck) appears in the center of a computer screen while simultaneously identifying the loca-tion of another target on the periphery of the screen.

Marsiske's results with his research subjects turned out to be even more impressive than he expected. Ten exercise ses-sions, each lasting sixty to seventy-five minutes, were all that his subjects needed to attain remarkable improvements mea-surable up to five years later, compared with people who didn't participate in the exercises. The memory-exercise performers did 75 percent better when memory-tested; reasoning-test participants did 40 percent better on reasoning exercises; and the speed-training veterans responded 300 percent faster. Taken together, a short investment in time and energy aimed at improving three areas of mental functioning led to sus-tained cognitive improvements and the prevention of age-associated declines in mental function.

What makes Marsiske's findings even more impressive was the age of the typical participant in the studies: an average of seventy-three years with a range extending from sixty-five to the early nineties. "But these findings apply to people in their fifties or even younger," he says. "This kind of training works no matter what your age. And the mental skills acquired ear-lier in life persist for many years."

Since building cognitive reserve is a lifetime enterprise, the earlier you start accumulating it, the better. That's because neuroscientists aren't certain when Alzheimer's dis-ease actually starts. For instance, children age seven to ten years who possess a gene (ApoE4) known to increase the risk of Alzheimer's disease show signs of reduced cognitive per-formance compared with children possessing other forms of the gene.

"This suggests there are cognitive differences very early in life," notes Jacob Raber, the Portland, Oregon, researcher who carried out the study. But, as Raber readily admitted when I asked him, ApoE4 isn't a thoroughly reliable marker for Alzheimer's disease even among aged adults: some older people with that gene show no signs of Alzheimer's disease either during life or at autopsy. Thus no one would—or should—make predictions that these children are necessarily destined to go on to develop Alzheimer's disease. And yet Raber's findings are troubling.

Ronald Reagan provides the most famous contemporary example of the difficulty in pinning down the earliest beginnings of Alzheimer's.

Did President Reagan's disease begin after he left office, or earlier? In 1987, the editors of the "Outlook" section of *The Washington Post* asked me to write an article addressing that question. At the time, toward the end of Reagan's second term, rumors were spreading rapidly in Washington that Reagan was suffering from Alzheimer's disease—based mostly on the vagueness of many of his responses during press conferences.

After reviewing transcripts and watching videos of Reagan's speeches and press conferences dating back to his days as governor of California, I concluded in my *Washington Post* article that Reagan showed no evidence of the disease while he was in office. The fact that doctors later diagnosed Reagan as suffering from Alzheimer's disease makes me less certain of that conclusion today. When I wrote my article, he might well have been showing the earliest signs of the disease. Yet pinning down exactly when President Reagan began showing indisputable signs of Alzheimer's disease isn't any easier twenty years later than it was when he was completing his term.

Throughout his adult life Reagan often couldn't come up with answers when asked specific questions. For instance, on February 5, 1962, Reagan, then fifty-one and president of the Screen Actors Guild, appeared before a Los Angeles federal grand jury called to probe antitrust allegations against MCA, the talent agency that represented Reagan. Throughout his testimony he offered responses as vague as those that would later characterize his news conferences during the later days of his administration. Was he afflicted with the earliest signs of Alzheimer's disease at fifty-one? Or did he simply possess a poor memory? Or did he do this on purpose?

As with President Reagan, the earliest signs of Alzheimer's often escape detection and are recognized as harbingers of the illness only after the disease has greatly advanced. Beyond sixty-five, the risk of dementia (largely but not exclusively due to Alzheimer's disease) doubles every five years. With statistics like these, anything that can be done to lessen the odds of Alzheimer's disease should be done.

Here are several other suggestions for lessening your risk for Alzheimer's disease:

Keep working for as long as you can at a challenging job that fully engages all of your talents and abilities. During your leisure hours follow the same principle: take up bridge, Go, or mahjongg. Get together with skilled Scrabble or poker players. When you're alone, read or do crossword or sudoku puzzles.

Establish and maintain a network of friends and acquaintances, since social isolation and the loneliness that usually accompanies it are now recognized as significant hazards to healthy brain function. To get a feeling for how important that is, imagine yourself mentally transported to a deserted island where there is no one to talk to and no means of sharing your experiences. Most people would find such a prospect extremely painful and distressing. That's because we're social creatures who have learned over the years that we *need* the presence of other people. Our penal practices provide a perverse recognition of this need: solitary confinement is reserved for those prisoners considered worthy of the harshest punishment. Health statistics also confirm the baleful effect of isolation. Lonely people are more prone to heart attacks, depression, suicide, and an assortment of physical and mental ills.

"I miss having people around me." "I often feel abandoned and completely alone." "I miss having a really good friend." These are some of the feelings expressed by lonely people on a test measuring loneliness. These comments describe an inner experience quite different from depression. A depressed person, typically, tends to socially withdraw; he simply cannot summon the energy needed for the give-and-take of social exchange. Lonely people, in contrast, crave human companionship and welcome any opportunity to mingle.

Recently, researchers have learned that lonely people are not only at greater risk for depression and certain other illnesses, but also more likely to suffer from cognitive problems, including an increased risk for developing Alzheimer's and other dementias.

According to a 2002 research study of twelve hundred elderly people in Sweden, the incidence of dementia, primarily Alzheimer's disease, was highest among those seniors with the least social contact. In general, the more socially involved the participants (frequent involvement in group activities such as sports, games, classes), the better they did. But even those with more limited socialization did better than loners, who tended to do things on their own and avoid other people. The Swedish findings were confirmed in 2007 by a study of eight hundred older people in the Chicago area that concentrated on loneliness. It found that lonely people are more than twice as likely to develop Alzheimer's than people who aren't lonely. Marriage serves as a protection against loneliness, but regular contact with close friends can be equally effective in combating loneliness. Whether one is married or single, it's important to counteract feelings of loneliness, since loneliness is a greater risk factor for Alzheimer's than depression or diminished physical and mental activity.

"Brain health and neural plasticity will be supported by anything that signals the brain that we are still an important member of the community and that adaptation is still required," according to Louis Cozolino, a psychologist at Pepperdine University in Malibu, California, and author of *The Healthy Aging Brain*.

In order to lessen the risk of loneliness-associated cognitive deficits, then, increase the time that you spend every day engaged in activities involving other people. On occasion you may have to force yourself to do this, especially if you tend to be shy or introverted. But as the research mentioned above indicates, you now have an excellent reason for making this effort: optimal brain health requires regular socializing.

In order to nail down the specific benefits of socialization I spoke to Dilip Jeste, director of the Stein Institute for Research on Aging, University of California, San Diego. He told me: "We have found several specific benefits of socialization: lower overall mortality rate, protection against hypertension and other cardiovascular diseases, less depression, higher cognitive performance, and most important, the delayed onset of dementia."

Think of it this way: The brain is a social organ that operates by the concerted activity of millions of neurons linked together by means of circuits—there is no such thing as a solitary neuron. Similarly, none of us exists in isolation; nor are we capable of optimal cognitive functioning unless—like a neuron within the brain's circuits—we become part of a wider social network. If you're experiencing loneliness, consider it your highest priority—for the sake of your brain—to take steps to bring yourself into meaningful social relationships. Join a hiking club, take a cooking class, volunteer at your church, recruit members for a book club—the choice will depend on your background and interests.

Learn ways to reduce mental stress. Mental stress is among the worst situations that you can create for your brain. I say "create" because mental stress differs from physical stress in an important way. While often you can do little about physical stresses caused by, say, a necessary surgical operation, you retain an important role in determining your level of mental stress.

Mental stress exerts its most harmful effect on the brain. The hippocampus diminishes in size—the explanation for why stressed people frequently complain about memory difficulties. Elevations of stress-related hormones such as growth hormone, epinephrine, and norepinephrine further decrease the ability to think in an organized manner.

Consider stress as one end of a mental energy continuum. For instance, think back to the last time you took a competitive examination. In order to perform at your best you had to muster a certain amount of mental energy (get "psyched"). If you couldn't do that and remained blasé about the test, you weren't likely to do very well. At the other extreme, if you became overly stressed, you also couldn't perform at your best—thanks to the harmful effects on your memory and general mental ability resulting from the surge of those stress hormones into your brain. This balance between too much and too little arousal forms the basis for the Yerkes-Dodson Law, named after the early-twentieth-century psychologists Robert Yerkes and J. D. Dodson.

A graphic representation of the Yerkes-Dodson Law (page 229) plots performance (vertical axis) against arousal (horizontal axis). The ascending slope of the inverted U-shaped curve represents the energizing effect of arousal. The downward slope, in contrast, depicts the negative effects of arousal (stress) on memory, attention, concentration, problem solving, and other cognitive activities. Notice that performance increases with arousal, but only up to a point (the height of the U-shaped curve). When the level of arousal increases beyond that point—as it does when a person is overstressed—performance decreases.

Yerkes and Dodson also suggested that there is an optimal level of arousal for different mental activities. While getting appropriately excited helps mobilize resources in a competi-

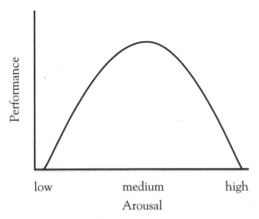

GRAPH OF YERKES-DODSON LAW

tive examination, a much lower level of arousal is best when writing a poem, painting a portrait, or meditating.

The best way to reduce stress is by mentally reformulating the stressful event or situation. Successful people do this all the time. "One person's stress is the successful person's challenge" is the key mantra they use to insulate themselves against the harmful effects of stress. In practical terms, this means the harmful effects of stress can be avoided by experiencing potentially stressful experiences as opportunities. Granted, some events will arouse stress in anyone (the death of a child, for example; or wartime experiences severe enough to induce post-traumatic stress disorder). But eventually even the most stressful situation comes to some resolution. And at that time you must make a decision that will determine how much stress this event will cause in your future life.

Ask yourself: "Am I going to allow this experience to determine the course of the rest of my life? Or am I going to take control of my life? What can I learn from this experience?" Mentally reformulating things in this way will increase your feeling of control and protect your brain from stress-induced damage. This is important, since loss of a sense of control is the main contributor to the stress response. No matter how stressful the situation, if you retain control of your attitudes and responses this alone will lessen your stress—even if you can't affect the situation responsible for your stress. Remember: even in situations where you cannot change what's happening to you, you can still change your attitude toward it. "Not being able to govern events, I govern myself," as the philosopher-essayist Michel de Montaigne described the process.

Practical exercises aimed at stress reduction often involve breathing. Unfortunately, most people breathe by expanding their chest and contracting their abdomen. This "chest breathing" is less efficient, since the lungs have to work harder to deliver sufficient oxygen throughout the body. This, in turn, results in a more rapid heart rate. And since only the upper and middle portions of the lungs are being recruited in chest breathing, optimal oxygen transport doesn't occur—leading to feelings of tension and stress.

To avoid these negative consequences, it's necessary to shift from shallow, predominantly chest breathing to deeper, more relaxing abdominal breathing. In order to do that, it isn't necessary to learn any of the specialized breathing techniques advocated by various schools of meditation. You're not seeking enlightenment here, but simply a reduction in stress.

Here is an easy method to shift from chest to abdominal breathing: Sit with your right hand on your abdomen and your left hand on your chest. Breathe deeply in such a way that the right hand rises and falls on your abdomen with each breath, while your left hand remains relatively still on your chest. Breathe in through your nose and out through your nose or mouth. After a few minutes spent establishing this abdominal breathing rhythm, you can forget about your hands and remove them from your chest and abdomen.

A second stress-reduction technique is to practice one of the mental exercises described in Part Three. In order to get away from repetitive stress-inducing thought patterns, choose one that doesn't involve reading, writing, or language in general. For example, quickly write out a random series of numbers and then commit them to memory using the zero-hero, one-sun, two-shoe, etc., mnemonic technique described on page 108. This imaging exercise (along with the other techniques described below) confers a double advantage: stress reduction and memory enhancement.

The simplest imaging technique involves nothing more complicated than looking at your immediate surroundings and taking a mental snapshot of them. Then close your eyes and "see" in your mind's eye as many details as you can. Finally, open your eyes and compare your internal snapshot with the scene in front of you. Note what you failed to record. Close your eyes again and add those things that you missed. Repeat the process until your mental image contains all of the relevant components.

Another way of practicing this exercise that I personally favor is to use a camera. First, I select a particularly appealing scene that I will enjoy conjuring at a later time when I'm under stress. Then I take a picture of it with the small digital camera I always carry with me. Later, when I'm feeling stressed, I envision (without looking at the picture) the scene in my mind with as much detail as I can muster. When I've done my best mentally recalling the scene, I then look at the picture and compare my remembered image with the photograph.

Other stress-reduction techniques include:

- *Learn to stop wasting mental energy on things that are beyond your control.* This doesn't mean thinking and acting as if you have no control over your life. As mentioned earlier, the worst stress results from situations in which we feel hopelessly dependent on circumstances we can do little about. But for most of us such circumstances are exceptions to the general rule that we remain in control of our lives—or at least we control the decisions that we make. In some cases, that means making changes in our work or personal lives. The most debilitating stress comes from "golden shackles" situations—you hate your job, but you have become dependent on the high salary that accompanies it; your marriage is miserable, but you stick with it because you don't want your children to grow up in a single-parent home. Such situations are even more stressful if you're unlikely to find a new job or remarry because of age or other factors. But even though you may require professional help to resolve such painful dilemmas, you have no other alternative than to reduce your stress by making a perhaps painful decision. Stress leads to a destructive cycle in the brain: depression and anxiety induce brain cell loss, which results in disturbances in thinking, concentration, memory, sleep, and general well-being. Taken together, these disturbances cycle back to produce even more stress.

- *Respond quickly when you can make a difference in resolving a stressful situation.* If you're feeling perpetually "keyed up" at work about a specific situation or in response to the behavior of another member of the work team, meet with him or her, explain your concerns, and try to work out your differences. If this doesn't prove successful, rethink your position. If you still find yourself experiencing stress because of the conflict between your views and the status quo, enlist the help of additional members of the team. But if you do that, be prepared for the possibility that the others may not see things your way. In such a situation, you have a decision to make: remain in the stressful situation or recognize that your best interests are served by moving on.

- *Work off stress with increased physical activity.* This will vary according to your health, interests, and external circumstances. Jogging, weight lifting, even walking will be fine. You don't have to join a health club—although that will provide the opportunity to vary your exercise routines and thereby avoid the biggest obstacle to maintaining a regular exercise routine: boredom. Take as an example the Japanese novelist Haruki Murakami, who lives as if each day were twenty-three hours long, so that no matter how busy he might be, he always has an hour to devote to exercise.

- *Avoid long, uninterrupted periods of work.* Take regular

breaks that engage your brain in a totally different activity. A lawyer friend of mine spends ten or fifteen minutes sketching during his breaks. This shifts his brain from verbal to visual—motor processing.

- *Monitor your moods, fantasies, and interior self-talk.* If you're feeling "keyed up," stop for a few minutes and figure out what stressors are contributing to your inner discomfort. If you find yourself dwelling on upsetting and stressful scenarios, change your brain activity by stopping whatever you're doing and turning to something that doesn't involve introspecting on personal concerns (exercise or working on a puzzle).

- *Reach out to other people.* With the construction and function of our brain, we're social creatures. As mentioned earlier, the brain is a social organ at every level, starting from the networking of its neurons and proceeding upward to our behavioral interaction with other people. Granted, this may have to be individualized according to your own personal "sociostat." If you tend to be introverted and uncomfortable in large social gatherings, it's all right to limit yourself to a few friends, but be sure to take the necessary steps to keep these friendships in good repair. The best way to do that is to take the initiative and make the first contact. Remind yourself of those occasions in the past when you felt better after getting together with someone, even though prior to the meeting you didn't "feel like" doing it. Your improvement in mood resulted from getting "out of yourself" and entering a less egocentric world.

- *As a corollary to the last suggestion, get away from stress-provoking thoughts by doing something for others.* Empathy and sympathy are two brain processes that are on the endangered list of human emotions thanks to constant media exposure to negative information and images (wars, terrorist attacks, genocidal tribal conflicts). The more of these images we see, the harder it is to put ourselves in the place of the people enduring these experiences (empathy) or to try to do something sympathetic to reduce their suffering. And while it's true that you are probably not in a position to do much about these larger concerns, you can be of service to people in your immediate environment. Volunteer at a soup kitchen, for instance, to enhance your brain's empathic faculties.

- *Try to reduce the importance of the event, situation, or person triggering your stress by keeping things in perspective.* If you're stressed in anticipation of an interview for a new job promotion, remind yourself that if this opportunity falls through, there will be others in the future. This technique, known as "reframing," enhances performance because fear and anxiety tend to diminish when the stressful situation is looked at with what the ancients referred to as the "long view." When taking the long view, you bring about activation of the frontal cortex, which then inhibits the stress-inducing activity of the amygdala, thus resetting your mental equilibrium.

- *Remind yourself that stress is the natural response to any situation that you can't manage or believe that you can't manage.* When you're feeling stressed, take the time to step back

mentally and ask yourself, "Am I setting myself up for failure here? Why not give it my best and see how things turn
out?"

- *Finally, remember that stress isn't all bad.* Your feeling of
stress is your brain's way of telling you that you must discover and address the cause of your stress. Failure to do so
leads to stress-related physical afflictions such as ulcers,
high blood pressure, and heart attacks. You can avoid
these afflictions by acting promptly and taking active steps
to change the stressful situation.

EPILOGUE:

THE TWENTY-FIRST-CENTURY BRAIN

A s I have emphasized throughout this book, healthy brain functioning involves a good diet, exercise, and, most of all, mental stimulation. The earlier you start on a program incorporating these factors into your life, the greater your cerebral reserve will be—your best bet when it comes to offsetting the effects of brain aging.

Here are my suggestions, based on discussions with renowned neuroscientists and my own observations. They are divided into four areas.

First, *nutrition:*

• *Reduce your cholesterol level either by diet or, if necessary, by taking a statin or one of the other cholesterol-lowering drugs.* The value of lowering cholesterol was illustrated by an imaging study (PET) on mentally fit older people. Those with higher cholesterol had decreased brain cell activity in those brain regions known to be at risk in Alzheimer's disease as well as areas impaired by normal aging. This suggests that high cholesterol plays an early role in tilting the brain toward an Alzheimer's pattern of brain cell loss. "Higher cholesterol levels conspire with other risk factors to trigger Alzheimer's," according to neuroscientist Eric Reiman of the Banner Alzheimer Institute in Phoenix, who carried out the imaging study on 117 healthy people in their fifties and sixties. Reiman's findings suggest that cholesterol-lowering treatments might reduce the risk of Alzheimer's disease.

  • *Try to include walnuts and blueberries in your daily diet.* Even though much of the research on the brain benefits of walnuts and blueberries involved animals rather than people, I believe the parallels with humans are sufficiently robust that you should probably include both of these in your daily diet. Walnuts contain an essential omega-3 fatty acid along with a host of other chemicals that act as powerful antioxidants to block the actions of free radicals that lead to inflammation and brain cell damage. The walnuts also inhibit the action of the enzyme acetylcholinesterase, which breaks down acetylcholine, the brain's key learning and memory-processing neurotransmitter. Eating between one and one and a half ounces of walnuts per day is sufficient to obtain the benefits, according to scientists at the U.S. Department of Agriculture Human Nutrition Research Center at Tufts University in Boston. A 2 percent extract of blueberries, or about half a cup per day, is considered sufficient to enhance brain circuits. If you don't like blueberries, strawberries do just as well, at least in animal research, in reversing age-related deficits in motor and cognitive behavior. Whichever fruit you select, the aim is to employ naturally occurring (and commercially available) compounds to halt or reverse some age-associated brain changes.

• *Consider increasing your daily intake of caffeine.* I say "consider" because not everyone can take caffeine. In some people it causes heartbeat irregularities (arrhythmias) and/or nervousness, sometimes called "coffee nerves." (I can't drink more than four or five cups of tea a day without feeling edgy.) Learn from your own past experience with caffeine and check with your doctor before increasing your caffeine intake. According to a recently initiated clinical trial measuring cognitive improvement on caffeine, memory impairment can be slowed and even reversed in mice given the human equivalent of five cups of coffee per day—about 500 milligrams. Caffeine isn't limited to coffee, of course, but can be taken in tea, sodas, and chocolate. But since not all of us can take caffeine, the best approach at this time is a cautious one. In the meantime, expect to hear a lot more over the next year or two about caffeine and its positive effects on brain function.

In order to improve *cognitive performance*, I suggest you try the following:

• *Avoid the principal factors shown to be associated with a decline in brain information-processing ability (cognitive decline):* insufficient physical and mental activity, a decreased number of friends, too much empty spare time, infrequent opportunities to converse, and excessive alcohol use.

• *Increase your capacity for sustained attention and concentration.* To do this, resist the pressures to multitask. In most cases, multitasking serves as a substitute for prioritizing. Even worse, it chisels away at your ability to remain mentally focused. And if you can't focus, your brain remains stunted and incapable of achieving its true potential.

• *Work on strengthening your memory.* Thanks to living in the most advanced technological society in human history, we rarely have to remember anything for more than the few moments it takes to enter that information into a PDA or PC. Since I'm always entering information into my digital recorder (random thoughts or on-the-spot interviews with people whom I've recognized as having something interesting to say), it's all the more important that I frequently exercise my own memory rather than use an electronic assistant. For instance, when I shop for groceries I don't read the items on my grocery list but challenge myself to remember them. Only when I'm ready to go to the checkout counter do I check the list for any omissions.

Learn memory exercises and perform them regularly. Specific suggestions can be found in this book for sensory memory, general memory, and working memory (see Part Three). Memory is our most vital mental faculty. It is also the first faculty to be affected by Alzheimer's disease. Strengthening memory is an important step in lessening the odds of developing that dreadful illness. As a neurologist I have personally observed improvements in my patients during the early stages of Alzheimer's among those who took active steps to improve their memory.

• *Increase your hand and finger dexterity.* Increased hand and finger dexterity has been shown to increase both mental agility and longevity. I suggest the game of Jenga as a fun tool for doing this; it combines manual dexterity with mental skills. It is played with fifty-four wooden blocks arranged so as to create an eighteen-story tower. Taking turns, each player removes one block at a time from any story except the top one. The removed block is then placed on the top of the tower. Additional blocks are then removed one by one from any of the lower seventeen stories and placed on the topmost story. The idea is to select blocks for removal and placement in a way that will not destabilize the tower and lead to its collapse. The game can be played either alone or with others.

For an additional benefit, combine finger-hand dexterity exercises like Jenga with juggling, which enhances the perception and spatial anticipation of moving objects—hence, an exercise that stimulates a wide swath of brain tissue that processes complex visual motion. What's more, the more skilled you become, the greater the resulting brain growth, according to the research of neuroscientist Bogdan Draganski of the Functional Imaging Library of University College London. Best of all, you don't have to achieve the skill level of a circus performer to see a benefit. Draganski found that the classic and easily learned three-ball cascade is sufficient. But once you learn to juggle, don't neglect your newfound skill. Spend a few minutes a day practicing to maintain the brain-enhancing advantages. In a Draganski study the brains of the volunteers who learned to juggle reverted to their pre-juggling pattern within three months of neglecting to practice a few minutes a day—a dramatic example that the brain's plasticity works both ways. Whatever activity induces a favorable change in brain structure or function must be continued to maintain that change.

• *Start playing video action games as a means of increasing your sensory sensitivity and rapid responsiveness.*

• *Buy and use a GPS, especially if you're over fifty-five.* Starting in middle age, the brain suffers difficulty correlating landmarks with the correct orientation and direction. Even when a relevant landmark is recognized, the person over fifty-five, for example, frequently turns right when he should be turning left. In response to the tendency to make such errors some people over fifty-five avoid long drives in unfamiliar surroundings. Part of the explanation for this deteriorating sense of direction is that as we grow older our hippocampus— the area of the brain concerned with navigation—activates less efficiently in situations where we're required to make decisions about direction. This decrease in proficiency isn't true just for our sense of direction; as we age our spatial sense in general tends to decline more than our language powers. You can counter this by using the GPS, but to get the most benefit, actively anticipate rather than passively respond: mentally construct a map linking where you are with where you're going. Do this even in familiar places. (Can you name all of the streets and roads you regularly encounter when driving from home to work?) The goal is to use the GPS not as a substitute for a failing sense of direction, but as a means of strengthening it.

• *Spend less time watching television.* A study published in the March 2006 issue of the journal *Neurology* compared the rates of cognitive impairment in a group of more than five thousand people fifty-five years and older. The researchers found that more time spent watching television was associated with a 20 percent increased risk of cognitive impairment. "Increasingly, it seems prudent to encourage persons of all ages, not just older persons, to turn the TV off," counsels an accompanying lead editorial. Not surprisingly, during the two months after the publication of the study, not many TV news outlets reported this research, which counseled viewers to turn off their sets because watching TV leads to a deterioration of their mental powers. Yet this is a message that sorely needs to be delivered in a society where the average school-age child spends twenty-seven hours a week watching TV. By the time that child reaches age seventy, he or she will have spent ten years of their lives watching TV. And by age seventy that lifetime of TV watching will be taking a cumulative toll on a brain already at great risk for cognitive impairment.

Why does watching TV exert such a stultifying influence on mental performance? It is not simply a matter of social isolation, or the noninteractive nature of television, or the concentration-jarring effects of repeated commercial interruptions, although each of these plays a part. Television induces changes in brain activity, shifting the balance slightly from the left hemisphere, which processes language and logic (organizing, judging, and analyzing information), to the right hemisphere, which favors impressionistic perception. Thus, emotional rather than analytical interpretations are more likely to come into play when watching a televised discussion about a topic than when reading about the same topic in a newspaper or magazine. In addition, television also induces passivity via an increase in alpha brain activity, a rhythm typically associated with unfocused receptive states of consciousness. We've all experienced this passivity. We become bored and tired of watching TV and yet we just can't summon the energy to turn it off. Most striking of all is our lack of intellectual engagement with TV as opposed to, say, the computer. As Steve Jobs puts it, "You watch television to turn your brain off, and you work on your computer when you want to turn your brain on."

• *Develop an appreciation of art.* Neuroscientists have recently discovered that different brain areas are activated by varied styles of painting such as Impressionism and Surrealism. Indeed, the brain's responses to the various styles are sufficiently distinctive that it's possible by brain imaging (fMRI) to distinguish whether a viewer is looking at a picture by Dalí or Picasso. This suggests that by increasing our familiarity with different schools of art we can exercise different brain circuits and enhance general brain activity.

• *Engage in reminiscence exercises.* For example, how many things can you remember that happened in 1994, or any other year of your choice? I perform this brain-stimulating exercise regularly. I jog my memory to come up with the major world events for a given year: who were the political candidates, the Oscar winners, the victorious teams in the World Series or the Super Bowl? Along with these details, I try to remember the specific circumstances of my life at that time. Each time I perform this exercise I find myself coming up with more details. That's because, as suggested by recent research by L.-H. Tsai of the Picower Institute for Learning and Memory at MIT, memory impairments may involve not actual loss, but rather failure on the part of the brain's neural networks to properly process memories. Based on her research, Tsai has concluded that new nerve cell networks can be activated to revive lost memories. Thus memories aren't permanently lost but potentially recoverable. "Past memories can be recovered even after a significant number of neurons have been lost in the brain," Tsai told me. This is a revolutionary and hopeful revision of traditional ideas about memory loss.

• *Develop a magnificent obsession.* I've observed that most successful people have obsessive character traits that they use to their advantage. That's the good news. The bad news is that these same traits also render them prone to negative obsessions: tendencies to ruminate, worry, fret, and kvetch about things that aren't under their control. One antidote for this is to develop what I refer to as a magnificent obsession: take up something that interests you but is far removed from your background, education, or life experience. Then start learning everything you can about the subject over the space of the next year. During periods of stress or whenever your mind drifts into negativity (worry, low moods, obsessions), shift mental gears from current preoccupations and devote a few moments to something related to your magnificent obsession.

If you're interested in World War II history, for instance, read every book that you can on the subject and make notes based on your reading; study film and video clips of the war; locate veterans in your area and interview them. Finally, collate all of this information in one document (I use a program called Microsoft OneNote). That way everything that you have accumulated can be accessed at one source. As it expands, read this document regularly. Finally, in order to keep yourself challenged, look for a program that creates new and creative rearrangement of the information you've gathered (Mindjet MindManager is one currently available program that can do this). At a certain point you'll be ready to contact experts in the field. While some experts may not welcome "amateurs," many of them, in my experience, are intrigued by a nonprofessional who achieves near-expert knowledge or performance in their field. In addition, join the relevant professional associations and attend their annual meetings. Each of these steps will increase your long-term memory storage. It will also help build up cerebral reserve.

Nor do the benefits depend on what subject you've chosen for your magnificent obsession. A magnificent obsession doesn't necessarily have to involve scholarly pursuits. An information specialist friend of mine became interested in carpentry several years ago and devotes some time every day to building furniture and working on home improvements. For complicated projects beyond his current abilities, he hires a professional carpenter and takes time off from his regular job to observe the pro and gain knowledge from him.

But whatever obsession you choose, keep in mind that different choices will engage different brain networks. Studying a historical period such as the Renaissance, for instance, engages the left hemisphere, primarily areas having to do with memory, reading, listening, imaging, and the pleasurable emotions that accompany learning new facts. My amateur carpenter friend is developing brain circuitry involved in form, scale, eye-hand coordination, and manual dexterity, along with the emotional satisfaction of creating something that can be seen, felt, and put to practical use.

One caveat: Since your intention is to enhance brain

function rather than achieve a level of *competitive* expert per-formance, too big a daily time commitment to your interest is likely to interfere with your job and your relationships. You can achieve most of the benefits from your project through regular, consistent efforts of no more than an hour to an hour and a half a day.

• *Every day find an hour to improve through practice a specific aspect of your performance in an activity that interests you.* If you're interested in golf, for instance, resist the impulse to go to a driving range and mindlessly hit buckets of golf balls. Instead, spend the time working on your greatest weakness—whether it be putting, chipping, or something else. "It is only human nature to practice what you can always do well, since it's a hell of a lot less work and a hell of a lot more fun. Sad to say, though, it doesn't do a lot to lower your handicap," said Sam Snead, widely considered one of the best golfers of the twentieth century. But practice isn't limited to athletics. Hobbies and special interests provide opportunities too. For instance, take up dancing. After a year of regular dancing once a week or more, you can expect improvements in your reaction time, fine-motor performance, posture and balance, attention, and nonverbal intelligence. These improvements not only are greater than what's found among people who do little or no exercise, but also exceed the brain benefits found among regular exercisers such as swimmers or those working out on exercise machines. Along with providing physical exercise, dancing gets us out of ourselves: we cooperatively engage with other people in learning the various dances and maneuvers; we improve our sense of rhythm and timing; we improve balance and coordination; finally, we counteract feelings of isolation.

In regard to *mood:*

• *Try to induce positive moods in yourself.* Longevity and general well-being have long been known to be favorably influenced by a positive mood. Maintaining a positive mood and attitude is especially important if you are sixty years of age or older, according to Dilip Jeste of the Stein Institute for Research on Aging. He has found decreased cardiovascular response to stress, along with a seven-and-a-half-year increase in longevity, among older people who remain upbeat and positive about their lives as they get older.

• *Use music to elevate your mood.* Research conducted over the past three years has shown that music elicits intense responses from the brain regions that process emotion, reward, motivation, and arousal. These are the same brain structures that become active in response to food and sex.

According to neuroscientists at the University of Zurich, it's possible to lift one's mood via a seventy-second exercise consisting of nothing more complicated than listening to a musical selection while looking at what the experimenters call "happiness" pictures (a man holding his smiling baby or laughing children playing by the ocean).

For example, in the emotional induction method developed by neuroscientists at the University of Zurich, the third movement of Beethoven's Sixth Symphony, when combined with a "happiness" picture, reliably lifts a person's mood. In each instance, the subjects were simply told to look at the pictures, listen to the music, and mentally place themselves into the same mood. In all instances the mood changes occurred within seventy seconds. Most important, their emotional experiences were most powerful in response to the pictures and the music together rather than either the music alone or the pictures alone. Combining the music with the pictures also induced the strongest activation in a distributed emotional network plus parts of the frontal, temporal, and occipital lobes of the brain. When you do this exercise, select musical selections with pictures that you find emotionally appealing. After combining pictures and music enough times, you'll find that you no longer need the pictures but can produce the same positive effect on your mood by mentally recreating the pictures in your imagination.

In another study on the effect of music on mood, carried out at the University of Pennsylvania, volunteers reported feeling happier after listening to selected music clips (a "jazzed-up" version of Bach's Brandenburg Concerto Number 3, for instance). While in this state of heightened mood the subjects performed better on memory tests than did participants in the experiment who didn't listen to mood-uplifting music. It's speculated that this memory improvement results from a positive mood-induced increase in attention. But whatever the explanation, the results suggest that you can use music to simultaneously lift your mood and increase your memory.

Obviously, the musical selections will vary from one person to another.

Discover the particular musical selections that arouse in you positive emotions under different circumstances. Think about your iPod musical library in a completely different way.

Why do I enjoy listening to Bach while home alone in the evening but enjoy Van Morrison while driving to work? By exploring questions like this you will discover the particular musical selections that arouse positive emotions under particular circumstances.

• *If you're prone to loneliness, combat it by scanning a few names from your address list and selecting one of them.* Then, after putting your relationship with that person into context in your own mind (a good friend, an acquaintance, business associate), consider calling the person and suggesting getting together for lunch or after-dinner drinks. Over the years I've often called writers, doctors, and lawyers I've known and suggested we meet. Even if the person can't get together immediately, I've felt less lonely after setting up a meeting for the near future.

• *Schedule regular naps during the workweek.* Naps not only improve your mood but also restore concentration and focus and increase creativity. I try to take a twenty-minute nap every afternoon.

Whether or not you're currently following any of the above suggestions, *it's never too late to enhance brain function*, according to a UCLA study involving subjects ranging in age from thirty-five to sixty-nine, with the mean age of fifty-three. The subjects in the study followed a program consisting of:

• A *diet high in omega-3 fats* from olive oil or fish, as well as fruits and vegetables rich in antioxidants (basically the Mediterranean diet, described on page 37).

• *Physical exercises* with the emphasis on cardiovascular conditioning such as brisk daily walks.

• *Mental exercises* aimed at strengthening memory and other cognitive functions. For instance, as a means of improving memory, the subjects learned to focus their attention by concentrating on, for later recall, random details of the clothing and accessories of family members. This was followed by other exercises aimed at improving visualization skills and the use of one of the mnemonic techniques as described in Part Three, combined with puzzles and brain teasers.

• *The practice of various relaxation techniques aimed at reducing stress.* After only fourteen days on this program the participants showed greater word fluency along with a decrease in activity in an area of the left prefrontal lobe associated with verbal fluency, working memory, and anxiety. It's speculated by the UCLA team that the memory and other mental exercises, combined with the relaxation techniques, caused the brain to function more efficiently with a decrease in the demand for glucose and other resources. The key finding, however, is the short time period in which these changes were brought about. Only fourteen days of a healthy diet, physical and mental exercise, and stress reduction induced dramatic improvements in cognitive efficiency in a part of the brain involved in memory and verbal fluency.

Although the UCLA study was a small one that enrolled only seventeen people, it has huge implications for brain enhancement. The brain, it's turning out, is even more plastic and malleable than the most enthusiastic researchers have heretofore imagined. Perhaps enhancement doesn't require months and even years of effort, but worthwhile alterations can be brought about in the brain in as short a time frame as two weeks (as occurred in the juggling study mentioned on page 241). Moreover, improvement doesn't entail expensive or inconvenient measures: no spas, health clubs, or specially prepackaged meals.

Most important, a brain-enhancement program confers benefits that can be discerned on testing five years later, as shown in the ACTIVE (Advanced Cognitive Training for Independent and Vital Elderly) trial conducted by the National Institutes of Health discussed on page 221. This study also involved a short time frame of ten sessions lasting a little over an hour. In the NIH study, the emphasis was on improving memory, mental processing speed, and reasoning.

Given all this, how should we think about the brain? What metaphor is most appropriate? Historically, many metaphors have been suggested. One of the more recent ones emphasizes the similarity between the brain and a computer. And while it's true that some of the brain's functions can be likened to the operations of a computer, this metaphor breaks down when it comes to physical structure and natural life cycles. Computers don't change in terms of their physical components; they continue to operate at maximal capacity until, at some unpredictable time in the future—and usually without warning, and under the most inconvenient circumstances—they abruptly cease to function. In many cases, after suitable fixes they can be restored to full function.

The brain, in contrast, changes in both size and function as it ages. The brain reaches its maximum size (measured by weight) somewhere between twenty and thirty years of age and decreases progressively for the remainder of its life span. Function too changes with age: as we age we experience decreases in reaction time, spatial processing, and working memory, among other functions. Yet these changes aren't due so much to brain cell loss—as was formerly believed—but to failures to maintain the neuronal circuitry linking neurons to one another. In support of this view, neuroscientists have found that the number of synapses linking neurons to one another in the cerebral cortex decreases with age, while the number of neurons themselves doesn't change very much.

Given these facts about the brain, the metaphor of brain-as-computer is of limited usefulness. What metaphor will enable us to put into practice all of the different pathways to brain enhancement and improvement suggested in this book?

The best and most helpful metaphor for the brain that I have come across was suggested by neurologist Kenneth Rockwood of Dalhousie University, Halifax, Nova Scotia, Canada. "Perhaps we should think of the metaphor of a series of marathons," he says. "As our brains age, we must prepare them to resist injury—equip them with good education, train them thoughtfully with challenging regimens, support them with nurturing environments, and be prepared to refresh them from time to time."

Rockwood's metaphor allows for both the structural and functional changes that accompany brain growth, development, and aging, as well as the active approach that will enable us to help our brain to achieve optimal performance throughout our lives. The metaphor also is consistent with the brain's varying performance depending upon its physical conditioning, which can always be improved by additional effort and training.

Rockwood concludes: "We should recognize that performance can vary dramatically from one marathon to the next. Perhaps the most important consequences of such a metaphor are that we should aim to see cognitive aging as a challenge for which we must prepare, that we gain enough equanimity to accept the slowing that even elite athletes experience, and that we reflect on what has been achieved along the way."

ACKNOWLEDGMENTS

Thanks to the following for their generosity in providing information, suggestions, guidelines, and, in many instances, richly rewarding conversations:

Carol A. Barnes, Ph.D., Regent's Professor of Psychology and Neurology at the University of Arizona, and research scientist at the Arizona Research Labs Division of Neural Systems, Memory and Aging.

Carl Cotman, Ph.D., director, Institute for Brain Aging and Dementia, University of California, Irvine.

Mahlon DeLong, M.D., professor of neurology, Emory University School of Medicine, Atlanta, Georgia.

K. Anders Ericsson, Ph.D., Conradi Eminent Scholar and Professor of Psychology, Florida State University.

Lorrie G. Foster, executive director, Councils and Research Working Groups, The Conference Board.

John Gregory Geake, Ph.D., professor of education at the Westminster Institute of Education, Oxford Brookes University.

Temple Grandin, author and assistant professor of animal sciences at Colorado State University.

Joshua A. Granek, Centre for Vision Research, York University, Toronto, Canada.

Kenneth M. Heilman, M.D., professor and director of the Department of Neurology, University of Florida College of Medicine.

Maya L. Henry, Ph.D., Department of Speech, Language and Hearing Sciences, University of Arizona, Tucson.

Dilip V. Jeste, M.D., director, Stein Institute for Research on Aging, University of California, San Diego.

Kate Karp, winner of the first annual National Adult Spelling Bee.

Arthur Kramer, Ph.D., Beckman Institute and Department of Psychology, University of Illinois, Urbana.

Michael Marsiske, Ph.D., Department of Clinical and Health Psychology, University of Florida.

Stephen Minger, Ph.D., director, Stem Cell Biology Laboratory, Wolfson Centre for Age-Related Diseases, King's College, London.

Martha Clare Morris, Sc.D., Rush Institute for Healthy Aging, Rush University Medical Center, Chicago.

Giulio Maria Pasinetti, M.D., professor of psychiatry and neuroscience, Mount Sinai School of Medicine, New York.

Michelle Pauls, AIMS Education Foundation, Fresno, California.

Jessica D. Payne, Ph.D., Department of Psychology, Harvard University.

Hal Prince, winner of the second annual National Adult Spelling Bee.

Clifford B. Saper, M.D., Ph.D., professor of neurology, Harvard University, and chairman of the Neurology Department, Beth Israel Deaconess Medical Center, Boston.

Gordon M. Shepherd, M.D., D.Phil., professor of neuroscience, Yale University.

Robert Strecker, Ph.D., Department of Psychology, Harvard University.

Matthew A. Tucker, Laboratory for Cognitive Neuroscience and Sleep, The City College of the City University of New York.

Ronald van Heertum, M.D., Department of Radiology, College of Physicians and Surgeons of Columbia University, New York.

Nora Volkow, M.D., director of the National Institute on Drug Abuse.

Matthew P. Walker, assistant professor of psychology, Harvard Medical School, and director of Sleep and Neuroimaging Laboratory, Deaconess Medical Center, Boston.

Dave Youngs, research fellow, AIMS Education Foundation, Fresno, California.

Special thanks to my editor, Jake Morrissey; my agent, Sterling Lord; my wife, Carolyn; and my three daughters, Jennifer, Alison, and Ann.

SUGGESTED READINGS

Part One. Discovering the Brain

Gazzaniga, Michael S., editor in chief. *The New Cognitive Neurosciences*. Bradford Books, The MIT Press, 2004.

Gluck, Mark, Eduardo Marcado, and Catherine Myers. *Learning and Memory: From Brain to Behavior*. Worth, 2007.

Purves, Dale, et al. *Neuroscience*. 4th ed. Sinauer, 2008.

Purves, Dale, et al. *Principles of Cognitive Neuroscience*. Sinauer, 2008.

Zigmond, Bloom, Landis, Roberts, and Squire. *Fundamental Neuroscience*. 2nd ed. Academic Press, 2002.

Part Two. Care and Feeding of the Brain: The Basics

Barberger-Gateau, P., et al. "Dietary Patterns and Risk of Dementia." *Neurology*, November 13, 2007.

Carroll, Linda. "To Sleep, Perchance to Cure." *Neurology Today*, July–August 2006.

Colcombe, Stanley J., et al. "Aerobic Exercise Training Increases

Brain Volume in Aging Humans." *Journal of Gerontology: Medical Sciences*, November 2006.

Korman, M. "Behavioral Interference Selectively Blocks the Evolution of Delayed Gains, but Can Be Countered by Daytime Sleep." *Society for Neuroscience*, 2006, program 366.21.

Kramer, Arthur. "Exercise, Cognition, and the Aging Brain." *Journal of Applied Physiology*, June 15, 2006.

Miller, Greg. "Hunting for Meaning After Midnight." *Science* 315 (March 9, 2007).

Morris, M. C. "Associations of Vegetable and Fruit Consumption with Age-Related Cognitive Change." *Neurology*, October 2, 2006.

Nishida, M. "Daytime Naps, Sleep Spindles, and Motor Memory Consolidation." *Society for Neuroscience*, 2006, program 506.7.

Phillips, Lisa. "A Mediterranean Diet Is Associated with Living Longer with Alzheimer Disease." *Neurology Today*, September 18, 2007.

Rubin, Jay. *Haruki Murakami and the Music of Words*. Vintage, 2003.

Scarmeas, N. "Mediterranean Diet and Risk for Alzheimer's Disease." *Annals of Neurology*, June 2006.

Sheth, B. R. "Practice Makes Imperfect: Restorative Effects of Sleep on Motor Learning." *Society for Neuroscience*, 2006, program 14-14.

Walker, Matthew P. "Sleep to Remember." *American Scientist* 94 (July–August 2006).

Walker, Matthew P., and Robert Stickgold. "Sleep, Memory, and Plasticity." *Annual Review of Psychology*, 2006.

Part Three. Specific Steps for Enhancing Your Brain's Performance

Baldo, Juliana V., et al. "Verbal and Design Fluency in Patients with Frontal Lobe Lesions." *Journal of the International Neuropsychological Society* 7 (2001).

Butterworth, Brian. "What Makes a Prodigy?" *Nature Neuroscience* 4 (2001).

"Chess for Drudges?" *Science Journal* 137 (September 21, 2007).

Ericsson, K. Anders. "Attaining Excellence Through Deliberate Practice: Insights from the Study of Expert Performance." In M. Ferrari, ed., *The Pursuit of Excellence in Education*. Erlbaum, 2002.

Ericsson, K. Anders. "Deliberate Practice and the Modifiability of Body and Mind: Toward a Science of the Structure and Acquisition of Expert and Elite Performance." *International Journal of Sports Psychology*, 2007.

Ericsson, K. Anders. "Exceptional Memorizers: Made, Not Born." *Trends in Cognitive Sciences* 7, no. 6 (June, 2003).

Ericsson, K. Anders. "The Path to Expert Golf Performance: Insights from the Masters on How to Improve Performance by Deliberate Practice." In P. R. Thomas, ed., *Optimizing Performance in Golf*. Academic Press, 2002.

Ericsson, K. Anders. "Protocol Analysis and Expert Thought: Concurrent Verbalizations of Thinking During Experts' Performance on Representative Tasks." In *The Cambridge Handbook of Expertise and Expert Performance*. Cambridge University Press, 2006.

Flynn, James R. *What Is Intelligence?* Cambridge University Press, 2007.

Hilts, Philip J. *Memory's Ghost: The Nature of Memory and the Strange Tale of Mr. M.* Simon & Schuster, 1985.

Kaul, P. "Psychomotor Vigilance Changes Following Meditation, Nap, Caffeine, or Exercise." *Society for Neuroscience,* 2006, program 361.2.

Kerr, C. E., et al. "Tactile Acuity in Tai Chi Practitioners." *Society for Neuroscience,* 2007, presentation 74.1.

Salthouse, T. A. "Effects of Age and Skill in Typing." *Journal of Experimental Psychology* 13 (1984).

Shur, Norman W. *2000 Most Challenging and Obscure Words.* Galahad, 1994.

Tammet, Daniel. *Born on a Blue Day: Inside the Extraordinary Mind of an Autistic Savant.* Free Press, 2006.

Walker, Matthew P. "The Role of Sleep in Human Memory Consolidation and Reconsolidation." In the mini-symposium "The Dynamic Nature of Memory." *Society for Neuroscience,* 2006.

Walker, Matthew P., and Robert Stickgold. "Sleep, Memory, and Plasticity." *Annual Review of Psychology,* 2006.

Part Four. Using Technology to Achieve a More Powerful Brain

Donath, Judith. "Virtually Trustworthy." *Science* 137 (July 6, 2007).

Granek, J. A., et al. "The Effects of Video Game Experience on the Cortical Networks for Increasingly Complex Visuomotor Tasks." *Society for Neuroscience,* 2007, program 8.24.

Graziano, Michael S. A. "Where Is My Arm? The Relative Role of Vision and Proprioception in the Neuronal Representation of Limb Position." *Proceedings of the National Academy of Sciences,* August 1999.

Green, C. Shawn, and Daphne Bavelier. "The Cognitive Neuroscience of Video Games." In P. Messaris and

L. Humphries, eds., *Digital Media: Transformations in Human Communication*. Peter Lang, 2006.

Larson, Christine. "Calisthenics for the Older Mind on the Home Computer." *The New York Times*, August 26, 2007.

Schiesel, Seth. "Another World Conquered by Video Games: Retirees." *The New York Times*, March 30, 2007.

Stein, Rob. "Real Hope in a Virtual World: Online Identities Leave Limitations Behind." *The Washington Post*, October 6, 2007.

Part Five. Fashioning the Creative Brain

Balzac, Fred. "Exploring the Brain's Role in Creativity." *Neuropsychiatry Review* 7, no. 5 (May 2006).

Bell, Madison Smartt. *Narrative Design: Working with Imagination, Craft, and Form*. W. W. Norton, 1997.

Book, David L. *Problems for Puzzle Busters*. Enigmatic Press, 1992.

Bottini, G., et al. "The Role of the Right Hemisphere in the Interpretation of Figurative Aspects of Language: A Positron Emission Tomography Activation Study." *Brain* 117 (1994).

Bowden, E. M., et al. "Aha! Insight Experience Correlates with Solution Activation in the Right Hemisphere." *Psychonomic Bulletin and Review* 10 (2003).

Bowden, E. M., et al. "New Approaches to Demystifying Insight." *Trends in Cognitive Science* 9 (2005).

Duch, W. "Brain-Inspired Conscious Computing Architecture." *Journal of Mind and Behavior* 26 (2003).

Eich, T. S. "fMRI Investigation of the Neural Bases of Deliberative Versus Intuitive Decisions." *Society for Neuroscience*, 2006, presentation 664.6.

Geake, J. G. "Functional Neural Correlates of Creative

Intelligence as Determined by Fluid Analogizing." *Society for Neuroscience*, 2006, presentation 664.1.

Jung-Beeman, Mark, et al. "Neural Activity When People Solve Verbal Problems with Insight." *PLoS Biology* 2, no. 4 (April 2004).

Mosley, Walter. *This Year You Wrote Your Novel*. Little, Brown, 2007.

Restak, Richard. "The Creativity of the Unconscious." In *The Brain Has a Mind of Its Own*. Harmony Books, 1991.

Sternberg, R. J, ed. *Handbook of Human Creativity*. Cambridge University Press, 1998.

Part Six. Impediments to Optimal Brain Function, and How to Compensate for Them

"Aging Population Is a Critical Business Issue." *Managing the Mature Work Force*. The Conference Board, 2005.

Draganski, Bogdan, et al. "Changes in Gray Matter Induced by Training: Newly Honed Juggling Skills Show Up as a Transient Feature on a Brain Imaging Scan." *Nature* 427 (January 22, 2004).

Ericsson, K. Anders. "Attaining Excellence Through Deliberate Practice: Insights from the Study of Expert Performance." In M. Ferrari, ed., *The Pursuit of Excellence in Education*. Erlbaum, 2002.

"Exercising to Keep Aging at Bay." *Nature Neuroscience* 10 (2007).

Jobe, Jared B., et al. "ACTIVE: A Cognitive Intervention Trial to Promote Independence in Older Adults." *Controlled Clinical Trials 22* (2001).

Lohr, Steve. "Slow Down, Brave Multitasker, and Don't Read This in Traffic." *The New York Times*, March 25, 2007.

Melton, Lisa. "Use It, Don't Lose It." *New Scientist*, December 17, 2005.

Raber, Jacob. "ApoE4 Affects Spatial Learning and Memory in Children." *Society for Neuroscience*, 2007, program 422.6.

Reiman, Eric. "Higher Midlife Cholesterol Levels Are Associated with Hypometabolism in Brain Regions Affected by Alzheimer's and Normal Aging." *Society for Neuroscience*, 2007.

Restak, Richard. *Older & Wiser: How to Maintain Peak Mental Ability for As Long As You Live*. Simon & Schuster, 1997.

Richtel, Matt. "Lost in E-Mail, Tech Firms Face Self-Made Beast." *The New York Times*, June 14, 2008.

Small, Gary W., et al. "Effects of a 14-Day Healthy Longevity Lifestyle Program on Cognition and Brain Function." *The American Journal of Geriatric Psychiatry*, June 2006.

Stern, Yaakov, et al. "Brain Networks Associated with Cognitive Reserve in Healthy Young and Old Adults." *Cerebral Cortex* 15 (April 2005).

Stone, Kathlyn. "Physical Fitness, Childhood IQ Affect Cognitive Reserve." *Neurology Review*, November 2006.

Valeo, Tom. "Neural Network Identified for Cognitive Reserve." *Neurology Today*, September 18, 2007.

Vedantam, Shankar. "Short Mental Workouts May Slow Decline of Aging Minds, Study Finds." *The Washington Post*, December 20, 2006.

Willis, Sherry L. "Longterm Effects of Cognitive Training on Everyday Functional Outcomes in Older Adults." *Journal of the American Medical Association*, December 21, 2006.

Wilson, R. S. "Relation of Cognitive Activity to Risk of Developing Alzheimer Disease." *Neurology*, November 2007.

Yamamura, H., et al. "Differences of Brain Activity Elicited by Different Styles of Art." *Society for Neuroscience*, 2006, paper 73.7.

Yamamura, H., et al. "Neural Decoding of Artworks: Can Brain Activity Tell Who Painted the Picture, Dalí or Picasso?" *Society for Neuroscience*, 2007, presentation 737.14.

Epilogue: The Twenty-first-Century Brain

Rockwood, Kenneth. "What Metaphor for the Aging Brain?" *Neurology*, November 13, 2007.

INDEX